When I Heard the Bell

THE LOSS OF THE IOLAIRE

John MacLeod

T0274435

BIRLINN

This edition first published in 2010 by
Birlinn Limited
West Newington House
10 Newington Road
Edinburgh
EH9 1QS

www.birlinn.co.uk

Reprinted 2013

ISBN: 978 1 83983 0 563

British Library Cataloguing-in-Publication Data
A catalogue record for this book is available from the British Library

Typeset by Brinnoven, Livingston
Printed and bound by Ashford Press, Gosport

for my father, Donald Macleod

Dòmhnall mac Dhòmhnaill 'ic Mhurchaidh 'ic
Dhòmhnaill 'ic Mhurchaidh Am Pìobair'

and the sailors of the Long Island

for my mother, Mary MacLean

Màiri nighean Mhurchaidh Chaluim Mhòir, mac Aonghais 'ic
Murchaidh 'ic Iain Òig 'ic Iain Duibh 'ic Iain mac an Abraich

and the women of Lewis

for the 284

and in honour of
Mrs Marion MacLeod (1914–2012)
last orphan of the *Iolaire*

Contents

Prologue

That was the night after the day after the longest night, the New Year night the children were up and down from the road, up and down, the table spread and the shift of clothes airing and all in readiness for a father scarcely remembered and a husband who would never, in fact, come on this New Year of peace; after the day the stuttering boy ran in, with the elders sombre at his heels, and she asked, 'Is it true?' and they said, 'Yes, it is true' – the night before the weeks before the night the cart came from Stornoway with its sealed, tarpaulined coffin, and inside only what was left and which none that knew him could look on . . . this was the night Dolina had her dream, and her drowned man came to her in the vision of her grief and amidst the rubble of her world and in the exhaustion of her mind, and she said to him, 'Oh, Iain, Iain, how am I going to manage?' And he said to her, his woman now widowed, 'Well, that's what I thought, too, when I heard the bell.'

Raoir Reubadh an *Iolaire*

'S binn sheinn i, a' chailin,
a-raoir ann an Leòdhas;
I fuine an arain
Le cridhe làn sòlais,
air choinneamh a leannain
Tha tighinn air fòrlach,
Tigh'nn dhachaidh thuic teàraint'
Fear a gràidh.

Tha 'n cogadh nis thairis
'S a' bhuaidh leis na fiùrain,
Tha nochd ri tigh'inn dhachaidh;
Tha 'n Iolaire gan giùlain.
Chuir mòine mun tein' i
'S an coire le bùirn air
Ghràidh, chadal cha tèidear
Gus an lò.

Bidh iadsan ri 'g aithris
's bidh sinne ri 'g èisteachd
Ri euchdanaibh bhalach
Na mara 's an fhèilidh;
'S na treun fhir a chailleadh,
A thuit is nach èirich:
O liuth'd fear deas, dìreach
Chaidh gu làr.

Cluinn osnaich na gaoithe!
O, cluinn oirre sèideadh!
'S ràn buairte na doimhne;
O 's mairg tha, mo chreubhag,
Aig muir leis an oidhch' seo
Cath ri muir beucach;
Sgaoil, Iolair, do sgiathaibh
'S greas lem ghràdh.

Ri 'g èirigh tha 'n là
'S ri tuiteam tha dòchas;
Air an t-slabhraidh tha 'n coire

Last Night the Iolaire was Wrecked

Sweetly she sang, the lass,
last night on Lewis,
as she baked the bread
with a heart full of joy,
expecting her sweetheart
coming on leave,
coming home to her safely,
her man beloved.

The war is now over
and the heroes have won,
and this night they'll be home;
the Eagle bearing them.
She put peat on the fire
and filled the kettle
my love, there'll be no sleeping
till daybreak.

They will surely relate
and we will surely listen
to the feats of the lads
of the sea and of the kilted soldiers;
and of the brave boys lost,
fallen, never to rise:
Oh, all those fine, upright men
who were struck down.

Listen to the moaning wind!
Oh, hear it blowing!
And the raging roar of the deep;
Oh woe's me, my calamity,
for those at sea tonight
battling the howling sea.
Spread, Eagle, your wings
and hasten with my love.

Rising is the day
and fading is hope;
on the hook the kettle

Ri pìobaireachd brònach;
Sguir i dhol chun an dorais
'S air an teine chuir mòine
Cluinn cruaidh fhead na gaoithe
A' caoidh, a' caoidh.

Goirt ghuil i, a' chailin,
Moch madainn a-màireach,
Nuair fhuair i san fheamainn
A leannan 's e bàite,
Gun bhrògan mu chasan
Mar chaidh air an t-snàmh e,
'N sin chrom agus phòg i
A bhilean fuar.

Raoir reubadh an Iolair
Bàit' fo sgiathaibh tha h-àlach;
O na Hearadh tha tuireadh
Gu ruig Nis nam fear bàna.
O nach tug thu dhuinn beò iad,

A chuain, thoir dhuinn bàit' iad,
'N sin ri do bheul clocrach
Cha bhi ar sùil.

Murchadh MacPhàrlain
Bàrd Mhealaboist
(1901–82)

pipes lamentation;
she stopped going to the door
and adding peat to the fire –
listen to the howl of the wind
grieving, grieving.

Sore she wept, the lass,
the morning of the morrow,
when she found in the wrack
her drowned beloved,
without shoes on his feet,
just as he had swum;
there she bent and kissed
his chill lips.

Last night the eagle was ravished,
her chicks drowned under her wings;
from Harris they mourn
to Ness of the blond boys.
Oh, if you could not give them to us
    living,
Ocean, give us them drowned,
then from your devouring mouth
we will expect no more.

Murdo MacFarlane
The Melbost Bard

Chapter One

The Dark Ship

'We have shared the incommunicable experience of war. We have felt, we still feel, the passion of life to its top . . . In our youths, our hearts were touched by fire.'

Oliver Wendell Holmes, Jr

On 31 December 1918, hours from the first New Year of peace since 1914, hundreds of Royal Naval Reservists from Lewis and Harris poured off trains from Inverness onto the pier at Kyle of Lochalsh.

The ratings on leave – 'libertymen' – were naturally eager to be home, as promised, with their families for New Year. Aware of this, the Rear-Admiral in command of the Stornoway Royal Naval Reserve (RNR) base had earlier that day sent their requisitioned depot-ship to Kyle to bear home all the Great War sailors the regular MacBrayne mailboat could not accommodate. At Kyle, the local Commander struggled with minimal information and a fraught problem of logistics. He cut corners. Dozens and dozens of men were marched aboard the ageing steam-yacht; so many most could not even secure a seat. Around 7.30 that evening, Tuesday, 31 December 1918, His Majesty's Yacht *Iolaire* set sail from Kyle to Stornoway.

She never made it. Just before two in the morning, only yards from the entrance to Stornoway harbour and but half a mile from a quay already jammed with eager friends and relatives, the *Iolaire* was wrecked on a notorious reef, the Beasts of Holm. Of the roughly 284 men on board – from middle-aged veterans to boys in their teens – only eighty survived. It is unlikely the precise numbers will ever be known. The disaster touched every family on Lewis and wasted entire villages, to say nothing of seven Harrismen who also perished. Of the ship's own crew, only seven hands were spared; all four officers were drowned.

The loss of the *Iolaire* is a tragedy without parallel, both in its mocking ironies – men who had come through global conflict to be washed up dead on their own doorstep, including very many who could themselves have safely steered her into port – and its appalling impact on one small, close-knit, fiercely patriotic community who had contributed prodigiously to Britain's effort in the Great War. The *Iolaire* remains not only Britain's worst peacetime disaster at sea since the sinking of the *Titanic* in April 1912, but the worst peacetime loss of a British ship in British waters in all the twentieth century. And no comparable event – such as the 1987 capsize of the Townsend Thorensen ferry, *Herald of Free Enterprise* – fell almost exclusively on one small, defined population.

There are only, perhaps, two events that come close to the wreck of HMY *Iolaire* for pathos, in the one instance, and evident moral turpitude, in the other. On 27 April 1865, hundreds and hundreds of half-starved but quietly rejoicing Federal prisoners, newly liberated from dreadful Confederate camps after the American Civil War, piled on board a stately side-wheeler, *Sultana*. A photograph survives, taken at Helena, Arkansas, of men cheering from her decks as she prepared to sail up river. Nineteen hours later, still on passage, her boiler blew up and some 1,700 men burned – or drowned – still many, many miles from home.

On 6 July 1988, the Piper Alpha platform detonated in the North Sea, below the eastern horizon from the coast of Aberdeenshire. Within minutes, all-consuming fire raged through the rig, as management dithered and all the agreed and supposedly rehearsed protocols of flight and survival proved useless. Of the 226 men on board, only fifty-nine survived, all of whom escaped by their individual wits and instinct, many jumping over 100 feet from the baking decks into the chill North Sea. Though vast compensation payments were with almost indecent speed arranged by Occidental Petroleum and its nonagenarian boss, Dr Armand Hammer, and the subsequent Cullen Inquiry uncovered both grave flaws in the rig's design and evident irregularities in her operation, the company was never fined and no one ever faced criminal charges.

This is but one point of striking similarity to the *Iolaire* disaster. In both cases, the economically vulnerable had been recruited into dangerous employment within a culture where safety concerns could not readily be raised, or authority easily questioned. On both the *Iolaire* and on Piper Alpha, the command structure on the spot disintegrated: no orders were given, no evacuation was directed and the men in charge focused obsessively on hopes of external rescue that never came. Ashore in Stornoway, floundering attempts to help from shore fixated on heavy, almost useless, life-saving equipment that only made the scene hours after the *Iolaire* sank. In the North Sea, a supposedly state-of-the-art fire-fighting vessel proved practically useless. The adjacent stand-by ship was ill-equipped for casualties, lacked basic medication for the injured, and had not even a working searchlight. A neighbouring platform continued for many minutes to pump gas into the inferno. At Stornoway, on 1 January 1919, the local commander likewise made no effort to get rescuers to the scene immediately and ascertain the facts. From both the blazing Piper Alpha and the collapsing *Iolaire*, the pitiably small bands of survivors made it entirely by their own efforts.

Seven sons of Harris, as we noted, were lost with the *Iolaire* – only for their names and their sacrifice to be obliterated from all public record for eighty years. Her dead crew left aching hearts all over the kingdom – Auchterarder, Grimsby, Hartlepool, Ipswich, Greenwich, Southampton, Suffolk, the Isle of Wight. But the rape of the *Iolaire* on the Beasts of Holm broke – ripped – the heart of Lewis and bequeathed a sorrow that has never healed. Nor will it, for decades yet to come.

Even today, ninety years after the catastrophe, the name of the *Iolaire* is still one uttered, on the streets of Stornoway and in the hamlets of South Lochs and the villages of Point and the grey, straggling townships of the West Side, with a conscious lowering of voice. Most native islanders over thirty can personally remember a survivor or a victim of first-degree bereavement – a widow, a parent, a son or daughter. Even today, there are at least two still living who were orphaned on 1 January 1919. The last widow went to her rest only in November 1980 – twenty years after the last mother of a drowned rating – and three survivors endured into the 1990s.

Exhibition on Lewis of a short art-house film on the disaster, in the autumn of 2005, caused distinct unease. In January 1999, when the local newspaper yet again – on the eightieth anniversary of the grounding –

published a comprehensive list of casualties, including even the *Iolaire* crew, but omitted the seven Harrismen, there was quiet fury south of the Clisham. There have been acres of verse composed on what befell the sons of Lewis on the Beasts of Holm – much sincere, most turgid; all but a handful quite failing to come to grips, either emotionally or intellectually, with what happened that night and just what it did to the Isle of Lewis.

The last of the quiet, marked band who had boarded HMY *Iolaire* and lived, Donald Murray of North Tolsta and Neil Nicolson of Lemreway, were buried only weeks apart in the early summer of 1992. Murray, Dòmhnall Brus – with an outstanding Great War record – could never discuss the events of that night without weeping. For decades, indeed, most who had made it ashore alive, soaked and chilled, bruised and bloodied and shaking, refused to discuss it at all. It was 1960 before a monument to the catastrophe was even erected at Holm Point (and some attacked the proposal); 1 January 1999 – years after the passing of Murray and Nicolson – before a memorial service was ever held. No one bothered to secure a Gaelic precentor for it, attendance was pitiable, and the wreath finally cast on the waters landed upside down.

These things indicate neither contempt nor carelessness. They testify, rather, to the depth of trauma – an event so apocalyptic that the day the *Iolaire* died could be recalled in intense detail by those who were only infants of three, four or five – recalling not merely a season that began with joy and ended with wailing, but life-defining social and economic disaster. The *Iolaire* widowed sixty-seven women, orphaned at least 209 children, and drowned six pairs of brothers. In two villages – Sheshader and Crowlista – none of their men aboard the ship survived.

And the stories of the days that followed are even worse – death on such a scale that the Isle of Lewis ran out of coffins; bodies recovered from the sea, weeks and even months after the grounding, in such a condition several could not be identified, in such a state that no one could bear to bring them home, and buried them instead at the nearest cemetery. In at least one instance – in a community where brave men of strong stomach were determined to retrieve their own – they reverently buried their burden near midnight, by flickering lantern, after the long,

long trundle by horse and cart over hill and moor, rather than let those immediately bereaved see and smell what the *Iolaire* had done.

In January 1919, James Shaw Grant was only eight years old, an English monoglot of a respectable United Free and self-consciously progressive Stornoway family. His father, William, owned and edited the newly launched local newspaper and stood at the centre of the biggest story of his career. 'It was the first event of my life which really got home to me,' James would write some sixty years later. 'I had no relatives, or friends, or even close acquaintances on board the *Iolaire*, but I still cannot speak of it, or even think of it, without being close to tears, as I have often been when I have read again my father's description of the scene, written hastily a few hours later . . .'

The younger Grant commented, too, on one nagging little irony. 'Iolaire' is Gaelic for 'eagle' – pronounced, in crude phonetics, YOO-luh-ruh. But, then and since, even on Lewis and even by the survivors, the name of the damned vessel has always been pronounced as it was by her incompetent officers: EYE-oh-lair. 'Chaidh a chall air an Eye-oh-lair,' they would mutter, the old men of my childhood, 'He was lost on the *Iolaire*.'

But there is another sad reality: furth of Lewis, over the Minch, HMY *Iolaire* and her catastrophic final voyage are almost completely unknown.

The world – insofar as it ever paid the tragedy much attention – has forgotten. In March 1987, when the *Herald of Free Enterprise* capsized as she left Zeebrugge, drowning 193 men, women and children, it was trumpeted on all sides as – yes – Britain's gravest maritime calamity in peace since the *Titanic*, a falsehood that infuriated Lewis. The sunken *Iolaire*, unaccountably, was not declared a war-grave – even though fifty-six bodies were never recovered – and there is hard evidence that, within a decade of her sinking, the hulk was largely salvaged for scrap.

I still, and with recollection of the eeriness of the moment, recall my first hearing of the *Iolaire*. It was July 1976, and I was a boy of ten. That summer is still remembered as a hot, searingly dry one for England, the summer of drought, of Mumsy tips to save water from Peter, John and Lesley on *Blue Peter*. In Scotland, though, the summer of 1976 was inordinately wet. It was a twilit evening in Stornoway – it must have been very late – and my father, at thirty-five rather younger than I am now, had taken the three of us for an amble round the harbour. (It must have been the night of the midweek prayer meeting, when a vacationing minister would be at pains not to be seen abroad while it was on, which is why it would have been late, for it was so dark the assorted harbour lights were on and the sky was fading from azure to indigo.)

We stood, obediently holding hands, near the head of No. 1 pier, and my father spoke of Arnish light, and Arnish beacon, and – yes – there had been wrecks; there was where the *Mamie* had grounded – and then, something tightening in his voice, he told us of the *Iolaire*, and all those men, so happy to be sailing home at last from the First World War, men who had braved U-boats and battleships and bombs and mines in the North Sea and in the Atlantic and by the Heligoland Bight, who had been drowned by the very gates of home – many of their bodies never even found for burial in their native villages.

Small boys have a ghoulish side. 'But why, Daddy, did divers not go down to the Eye-oh-lair and get the bodies?'

'John, John. Who on earth would want to dive deep into cold water and go about a sunken wreck among a lot of dead and rotting bodies?'

My forensic streak was also well advanced. 'And whose fault was it that the boat went on the rocks?'

'It was New Year and the officers were drunk.'

We drove back to my grandparent's cosy house in Newmarket, high on the Barvas road and with a view over the Cockle Ebb and Broad Bay and Point, with the big church at Back in sight and the flashing lighthouses of Tiumpan Head, and, much further away, Stoer Point and Cape Wrath. Family worship was imminent; the conclusion of which was cue for implacable parents to usher, tenderly but in tones that brooked no resistance, holidaying little boys to bed.

I determined to delay family worship. 'Grandpa, do you remember the Eye-oh-lair?'

My grandfather, my *seanair*, still only in his mid-sixties, was a slight, spare man with a cloud of white hair and vivid eyes, prematurely retired, his health broken by the hardships of naval service during Hitler's war.

Something flickered over his face – a darkening, awful recollection, passing over his features as a cloud in summer trails a shadow over the island moor.

He answered, not to me but to my father, and not in English but in Gaelic. My father asked questions. My grandmother spoke too, and for some minutes they conversed, entirely in Gaelic and in tones of palpable solemnity. Nothing was said to us, and it was understood, without intimation and without resistance, that this was something not, for the moment, to be spoken of in the understanding of small children. Then the Bible was lifted, and worship began.

That October, though, my grandfather came down to Glasgow for a brief vacation in our manse, and of a Saturday evening we were propelled into the sitting-room to be alone with him and briefed to ask him about the old days. I asked him again about the *Iolaire*.

Without my parents about, without his wife, Seanair spoke freely, earnestly. 'I remember the carts,' he said, 'the carts coming to Ness with the coffins. So many coffins – two coffins, four coffins, or six to a cart . . .' And he was not merely remembering the coffins in that moment – he was *seeing* them, his mind far away, in the boots of an eight-year-old boy near Cross Free Church, standing as the carts clopped by for Swainbost and Habost, Lionel and Skigersta, Knockaird and Eorodale and Port of Ness, coffins and coffins and coffins. And why? 'It was New Year. The officers had been drinking.'

And he said, 'My father, your great-grandfather, could have been on her, but at the last minute they put him on the *Sheila* instead, because he was standing with the boys on the left . . . '

The following summer, that of 1977 and Silver Jubilee, my father took me fishing by Holm, under strict instruction that I was to sit just where he put me and on no account to move. There was certainly no question of me fishing, inches from the edge of a rocky ledge over green, oily, seemingly bottomless swell, where dark kelp swayed like the arms of dead men. The inscribed granite memorial sat by the sprawled cairn that had long preceded it and, *air Biastan Thuilm*, on the Beasts of Holm, the tall, lightless beacon – it was still red in those days – rose skyward in silent rebuke. Only months earlier, a local fishing boat had foundered on this reef. We had seen her half-salvaged hulk by the Holm shore; by the Saturday of Stornoway Carnival, she had been floated far into the tidal reaches of Stornoway harbour, resting against the quayside, already barnacled and with a jagged hole in her planks. Two men had died.

I sat forlornly, too scared to move, watching the sea and the suck of slack, thick tide, my father landing mackerel, and the bared, wracked rocks of the Beasts of Holm, and I thought it – as I think now – a dark place, a place one might visit for momentary homage, not one where any who knew what had happened there would long linger.

Holm has scarcely changed since January 1919, and not significantly changed since my boyhood, save that the disaster has now – ever so delicately – become part of the market commodity that is West Highland history. In 2003, a signpost was erected for the HMY *Iolaire* memorial, leading you off by a narrow rightward road from the main highway to Stornoway Airport and the villages of the Eye Peninsula, and the grey coastguard station at the end of this road had been long demolished. (A new and much better facility finally replaced it, by Goat Island in Stornoway Harbour.)

Today, the road has been extended further over the pastures of Stoneyfield Farm, and 2002 brought the dreaded 'interpretative facilities' – a plaque, in English, with its mawkish 1919 cartoon of a grieving (and distinctly butch) island woman from the *Stornoway Gazette*, and with one or two serious inaccuracies. A footpath, too, was laid all the way to the monument, for the presumable benefit of the disabled, though the gravel is coarse and such a highway should not be lightly essayed with a wheelchair.

In the summer of 1988, walking out from Newmarket – in the days when the hills and inclines of island roads felt far less steep than they do now – I had been disturbed to find, near the seventieth anniversary of the sinking, how badly weathered the lettering on the *Iolaire* monument now was. I had tramped back, some days later, after calling at Stornoway's legendary hardware store, Charles Morrison Ltd, with its mighty counter and big drum of paraffin and fishing-tackle (long closed, now, succeeded by a trendy restaurant) to procure a little pot of the appropriate black paint, a brush, some white spirit and a rag.

I crouched for as long as it took carefully to repaint each embossed letter. I had just graduated from university. I had long read up on the tragedy and, especially, read what I could of the only definitive history, the outstanding *Call na h-Iolaire* ['*The Loss of the Iolaire*'] by Tormod Calum Dòmhnallach, largely in Gaelic but with maps, charts, an English synopsis and a navigational appendix. I now knew that the subsequent Fatal Accident Inquiry at Stornoway had indignantly repudiated any suggestion that the officers of HMY *Iolaire* were the worse for drink – and I knew that nevertheless remained the private conviction of an entire generation, a generation already passing away from the realm of time and space.

Through my teens and into manhood, I spent more and more hours delving into island history, into matters maritime, talking with my grandparents, with great-aunts and grand-uncles and other elderly tradition-bearers, asking about many things and in particular about the *Iolaire*. Later that year, I embarked on journalism, by the kindness of BBC Highland in Inverness. My radio career proved brief – my voice, then thin and high and still, today, inclined to rapidity of utterance, is not that of the born broadcaster. But that December of 1988, I was encouraged to produce a little English documentary on the *Iolaire* under the supervision of the long-suffering Angus MacDonald (though, wisely, he insisted on presenting it himself) for broadcast on Radio Highland in January 1989 to mark the seventieth anniversary. By then only three survivors were spared and only one, the gentle Donald Morrison of Knockaird, Am Patch, who had the most astounding escape of all, was really fit for interview.

I used recorded soundtrack from an interview Mr Morrison had granted Andy Webb, the assured and blandly handsome young Englishman then retained as the BBC's solitary television reporter in the Highlands, and months earlier – the only occasion I ever did so – I had personally met an *Iolaire* man, Donald Murray of 37 North Tolsta.

Our time was constrained and the circumstances somewhat chaotic, but these were precious minutes. Though now ninety-three and largely housebound, Dòmhnall Brus was still of powerful physical presence: chest like a barrel, vast hands, surprisingly dark hair, eyes like thoughtful pools. He was most affectionate and tactile, one who grasped your hand and stroked your thigh, who called a youth '*a ghràidh*' in the unselfconscious manner of his generation. We spoke largely of easy things: his mighty prowess in youth as a swimmer – he was in a Navy water-polo team – and his Great War service.

Murray served for the entire First World War, alongside his best friend John Morrison (Iain mac Choinnich mhic Iain Mhoireastain) and they were almost captured in Holland when, like very many Royal Navy men, he was drafted into a half-cocked scheme by Churchill to form a sort of sailor's army, the Royal Naval Division, and dash for Antwerp to defend it from the Germans. (The Division was not a success. Sailoring and soldiering are clean different trades, calling for very distinct skills; the Naval men did not prosper in the trenches of the Western Front.)

The mission failed, thousands fled into Holland, and were imprisoned for the rest of the war by that strange little land's complacent policy of neutrality. The prisoners included many island men who were even allowed to go home each year for a few weeks – usually for springwork or harvest – but had dutifully to return to Dutch imprisonment and a diet, as one Lewis scribe wrote darkly at the time, of 'black bread and horseflesh'. Britain could not provoke Holland into alliance with the Kaiser – further extending his control of Channel ports – but there were few wet eyes on Lewis when, in the summer of 1940, the Netherlands learned the hard way that neutrality is seldom prudent and never admirable.

Donald and John, though, dodged capture with exuberant resourcefulness. Their war continued. They were among the first on the scene when the *Lusitania* was torpedoed, a 1915 'atrocity' still the stuff of great controversy but an act which helped turn American public opinion towards Great War intervention. They were later at the Dardanelles, by the bloody shores of Gallipoli, and all over the Mediterranean. And they were together on the *Iolaire*, and there Donald survived and John was drowned. 'They couldn't get out,' said Murray, 'the men in that saloon, they couldn't get out . . .', and his hands rose and he began to weep, and we spoke of the *Iolaire* no more. 'I went through great experiences . . .' he mused before we parted – things wonderful, things terrible. I did not again intrude on his privacy, but I have never forgotten the depth of Donald Murray's emotion, the strength of his thankfulness or the breadth of his sorrow.

'There were seventy-nine survivors,' noted the journal of the local Tolsta historical society in 2006, 'but for years they could not speak about their experiences. It was said that the relatives of the survivors and the survivors themselves, although grateful, almost felt guilty that they survived.'

One village boy – a young Christian, a Free Presbyterian communicant who had already scrambled to safety, swam back from shore to the stricken ship because he could not find his brother – and so it passed that both were drowned. An older brother from Swainbost, a Ness township, likewise abandoned land to return on a like, frantic errand; these lads, too, were lost.

But such men, like all those Great War dead, were fast idealised by their community, remembered by their friends, cherished by grieving parents and somehow made sinless, perfect, forever young. Survivors of the *Iolaire* lived on in these same villages, amidst unspoken tensions and emotion, and for many a year it must have seemed they might – to steal a line from a *Titanic* melodrama – wait always for an absolution that would never come.

Two of the Tolsta survivors – there were only five, out of the sixteen who had boarded at Kyle – left the village permanently. Nor was the *Iolaire* any inoculation against future affliction. In his middle years – and at real danger to himself – Donald Murray swam out to retrieve the body of a young son, drowned off the Tolsta rocks. Donald Morrison married in 1937 and was widowed just two years later; in the 1950s, his Knockaird home was twice blasted, and badly damaged, by lightning.

As the war against the Kaiser passes remorselessly from living memory – in 2008, the last French and German veterans have died – there has been recent and intelligent reassessment of the conflict, going some way to correct the damage done over decades by British self-loathing and a few famous middle-class poets.

It is important to set the loss of HMY *Iolaire* in the context of the Lewis society of its day and of a community, an outlook and a way of life now, in most respects, as alien to us as if it were another planet. And grasp the sheer toughness of a people used to grinding physical toil, who walked prodigious distances even in old age, who memorised complex genealogies, catechisms, poems and Scriptures, who slaughtered their own cattle, sheep and fowl, who washed and dressed and buried their own dead, who lived vigorously on a scant plain diet and had – amidst discomfort and griefs beyond our comprehension – the most enormous fun.

It must be placed, besides, within wider understanding of the Great War and some grasp of its meta-narrative.

We shall look shortly and in the specific experience of one Lewis village at how men and women, boys and girls, got on with their lives during the First World War, if only to marvel at their resilience, their serenity and their joys amidst untold pressures – and that not only from war, battle and shortages, but from the dangers of their daily environment and from diseases then common, incurable and too often deadly.

But we need, besides, to start unlearning much of what we think we know about the Great War itself. We tend to think it was completely unnecessary – waged for class-ridden imperial splendour; that it was incompetently directed; that not a tenth who served survived; that untold thousands of shell-shocked men were shot as deserters; that the officers were upper-class twits who lounged over breakfast even as the plebs they commanded went over the top; that soldiers on the Western Front spent four solid years, without respite, in the slimy and shell-blown trenches; that Britain 'lost a generation'; that every household in the land lost a relative.

Unfortunately, it is all completely untrue – notions in a new land where no one much under seventy has any clear memory of a country in total war; myths long cosseted in a society which cannot imagine the hardiness of great-grandparents or the order of their world.

Britain could not dodge the Great War. For one, we had explicit treaty obligations to Belgium and clear understandings, in the 'Triple Entente', with France and Russia. Imperial Germany was, besides, a ferociously ambitious and dangerous state. Apart from the outrage of her borders in the Schlieffen Plan, the Belgians endured real atrocities, well documented. And from the mid nineteenth century, as the Prussian monarchy drove towards German unification, there was a sustained pattern of aggressive war – annexing Schleswig–Holstein in 1864; against Prussia's erstwhile Austrian ally, in 1866 (giving Berlin in victory the chance to annex besides four other German states); and against France in 1870. The following year, even as the siege of Paris continued, Wilhelm I, king of Prussia, was declared German Emperor at Versailles. Thereafter, Germany continued rapidly to industrialise, to acquire colonies in Africa (even by the standards of the time, they treated the natives abominably) and, though notionally a democracy, with strong education and public health provision, she was in truth an utterly militaristic state. Generals were far more important than politicians, and the elected Reichstag had no authority over foreign policy, the army, the navy, the conduct of war and the negotiation of peace: these were all the prerogative of the Crown.

In 1888 the old Emperor died. His son Frederick – educated, urbane, and of liberal instincts, a son-in-law of Queen Victoria – was himself gravely ill. His reign lasted but ninety days and he was succeeded by his son, another Wilhelm, only thirty and a man who had been scarcely on speaking terms with his parents. Wilhelm II was a profoundly damaged young man: bombastic, spiteful, inadequate and vain. Within two years he dismissed the great Chancellor, Otto von Bismarck, who had masterminded German unification while pursuing the shrewdest foreign policy. His successors in government had neither his wisdom and foresight, nor any significant influence over the brash young Emperor. Germany continued to pursue expansionist, reckless policy, soon alienating Russia as well as France, and by 1914 – by which time, not least by the Kaiser's antics, she had forfeited British regard and trust besides – she was surrounded by potential enemies and with no significant allies. At least as early as 1910 she was determined on war with all three powers and planning explicitly for battle with Britain. Even in terms of our own immediate self-interest – to say nothing of moral obligation to others – Britain could not have allowed such a despotism to emerge as the untrammelled master of Europe and, acutely, in control of the Channel ports and the Rhine delta, threatening Britain herself and the Royal Navy on which her security and commerce depended.

The war did not go well. As late as July 1918, the outcome was in genuine doubt. Germany had already beaten Russia (now in the grip of

Bolshevik revolution) and forced her into vast territorial concessions. We were, besides, but a junior coalition partner, with key decisions being made by the French. The awful Battle of the Somme was waged expressly to relieve them from crisis at Verdun – just as Third Ypres, waged a year later, was again to cover France, her exhausted army being then in a state of mutinous collapse.

In 1914, tactics had not caught up with new technology. (All the necessary lessons were there from the American Civil War, had an arrogant Europe bothered to study it.) Generals still believed that to mass your fire you had to mass your men; politicians to think that war, entrenched, could be waged to success by grueling attrition, wearing down not merely your foe's fighting strength but his economy and his morale. But the real problem was our own very small standing army and the long, long learning curve necessary to train up an enormous new one, raised first from volunteers and later by conscription. There were some undoubted mistakes at the Somme – chiefly the complacent assumption that a sustained, preceding artillery barrage would destroy both German positions and the miles of barbed wire – but the real reason why casualty figures were so frightful was the inexperience of most troops and a grave shortage of battle-hardened officers. The real reason it lodged so deeply in the national consciousness was the ill-judged system of territorial recruitment, since abandoned: if one district company or 'pals' battalion' was in a hot spot, almost every man in a village or factory was killed.

Too much is made by armchair historians of the First World War of, for instance, the significance of the machine-gun. Barbed wire – which put paid permanently to the cavalry charge and was largely immune to artillery barrage – was much more decisive a factor in the protracted deadlock; the deadliest weapon, the high explosive shell. Nor – despite much more modest casualties – should the achievements of the Royal Navy be forgotten. The Germans never won any mastery at sea, and were ruthlessly blockaded into final collapse – and the later, enormous movement of men and materiel from the United States owed all to the prowess of British ships and the men who sailed them.

Despite real, early disadvantages, Britain learned fast on every front of fighting endeavour, and the final victory was a triumph of which we might, today, be justly proud; the achievements of our army in 1918 are, by any fair criteria, its greatest in history. The casualties, of course, were dreadful. In the Great War, 744,702 British men died; in the Second World War only 264,000. A glance at any village memorial underlines the vaster human cost of the First.

But soldiers did not die by the thousand because of military incompetence and irresponsible commanders: as 'lions led by donkeys', to use an infamous quote of doubtful provenance. They died because, for over four years, in ceaseless fighting, Britain met day upon day the massed armies of her principal enemy on one murderous front. In the conflict of 1939 to 1945, by contrast, we met such intense land war only in the final year. Where the Allies then fought in comparable conditions – the siege of Bastogne; the Bittburg campaign – casualties were scarcely less frightful. We have quite forgotten the appalling price paid by the Soviet Union – 20 million died in their four-year war of attrition – and we are startled to learn that, at Normandy in June 1944, a higher proportion of British soldiers were killed than in the whole Battle of the Somme. The men of the Great War had, besides, to face new, lethal technologies of death with neither reliable motor transport nor portable battlefield radio, making it far harder to gain advantage and break from entrenched fighting to a war of sustained movement. Struggle of this intensity will always cost the lives of very many young soldiers; as General Mangin, a French commander of the Great War, had to declare, 'Whatever you do, you lose a lot of men.'

There was no 'lost generation'. Most British men who fought in the Great War survived and our casualties were markedly fewer than those of France or Germany. Officers did head into battle with their men: of all British servicemen involved, 12 per cent of 'other ranks' were killed, but a murderous 17 per cent of all officers. Over 200 British generals were slain or captured.

Of the five million who served, only 2,300 were sentenced to death by military courts between 1914 and 1918 – and 90 per cent of them were pardoned. Only one British household in fourteen lost a family member. Speaking personally, my three surviving great-grandfathers and my Shawbost grandfather not only served throughout and outlived the Armistice by very many years, but had buoyant memories of the experience, shadowed starkly – of course – by the *Iolaire*. We lost only one near kinsman, my Habost great-grandfather's younger brother, another Angus Thomson, who drowned in February 1918 with HMS *Eleanor*.

It may be years before real history by the likes of Norman Stone, Gordon Corrigan and Gary Sheffield can undo the impression bequeathed by untold English teachers. But, in 1914, our declaration of war on imperial Germany was hailed enthusiastically by the vast majority of British workers, by the trade unions, and by sixteen out of Glasgow's eighteen Independent Labour Party councillors. Scots rushed to volunteer. When Earl Haig – so vilified by Blackadder – died in 1928, his body lay in reverent Edinburgh state; 100,000 people filed past his coffin.

The disaster was not that the Great War was fought, but that it was not completed. Germany should have been invaded and occupied after the fourth Prussian essay of aggressive war in a lifetime; or at least met one humiliating, indubitable defeat on her own soil. Instead her soldiers were allowed to retreat under armistice from foreign land she had criminally occupied, in good order and bearing all their arms, and – as both Haig and Pershing warned at the time – were not slow to spread the potent myth that German soldiers, but for political chicanery at home, would have certainly triumphed. The terms imposed on Germany at Versailles – which legally concluded the Great War on 28 June 1919 – may indeed have dangerously humiliated her, without in like prudence actually weakening in any significant way her ability to make brutal war again. However, in the political and popular realities of the time, it is difficult to see what other terms the Associated Powers could have agreed.

Scotland's tragedy was not the War, but the War's aftermath. With the Russian Revolution, we lost our chief market for cured herring, and the fishing industry collapsed. Postwar politicians, of whatever party, utterly mismanaged the economy at profiteers' behest. Interest rates were high and this discouraged new enterprise – making things people actually wanted to buy. Massive debt to America brought high taxes and minimal welfare.

Soon, there was emigration. Less widely remembered is the huge internal migration – from the Hebrides, from the Borders, from Assynt and the Mearns – to towns and cities. Our Scottish countryside has never been replenished; the lively, Highland, Gaelic-speaking townships of 1914 are today's holiday-let deserts, or bastions of English colonialism. The war, as Donald MacCormick and others have rightly pointed out, was – as a protracted social event – a calamity for Gaeldom. The absence of so many men for so long permanently weakened Highland agriculture in the most marginal districts, and the general upheaval and loss and the economic difficulties that quickly beset the 'land fit for heroes' initiated massive movement of people – not so much over the Atlantic and to the colonies, but from Assynt and Torridon, Lochaber and Moidart, and Skye and Harris to the industrial lowlands, the coastal shipping trade and especially to Glasgow, and not just young men but, in very large numbers, young women.

It could justly be said that in 1914, Scotland was the richest realm in the Empire – if wealth based on untrammeled free-market enterprise never in the least subjected to intelligent Christian critique, and heaped high at frightful social cost. It was built largely on the production of raw materials – coal, steel, shale-oil – skilled heavy industry (the building of ships, the manufacture of locomotives) and textiles. We should not underestimate the might of Scottish agriculture, her eminence in a host of academic and scientific fields, or the surprising vigour of her publishing sector. A decade later, Scotland was a fiscal basket-case and has continued, really, to flounder ever since, increasingly dependent on public sector employment and far too complacent about a new, modern underclass, benighted by dreadful health, low life-expectancy, entire spiritual poverty, multi-generational unemployment and despair.

And it could be said, with equal accuracy, that in 1914 the Isle of Lewis was the richest part of the West Highlands. Stornoway supported a strong retail economy, a thriving cultural scene, excellent amenities and a considerable herring fishery, all set about a magnificent and most sheltered harbour.

A new industry, the production of Harris Tweed – not then the lure of rich tourists, but woven by the mile for the thousands retained on Edwardian sporting estates – had been deftly commandeered by Stornoway merchants. The 1911 census recorded the highest island population we have to date seen – 29,603 – and not one imbalanced, as today, by a preponderance of the elderly. Raised on a life of hard outdoor work and an elemental diet of potatoes and oatmeal, fish and milk and butter, young Lewismen and women were remarkable for their vitality, with glowing complexions and perfect teeth. Part of the enduring hurt, gazing on the photographs of the many lost on the *Iolaire*, is just how handsome so many of these lads were.

And we should abandon notions of meek Highland fatalism on this island. Though harried at times by the odd nasty factor or incompetent official, there had been minimal Clearances on Lewis – nothing on the scale of what befell the Uists, Harris and Skye in the early nineteenth century. The Lewis folk, besides, with a certain assertiveness that came, perhaps, from a very strong dose of Viking blood and the ancient stability of their villages, were not lightly messed about.

We are so inured to a narrative of passive, saintly Highland losers trooping obediently onto emigrant ships that it is startling in the nineteenth century to see repeated flashes of undoubted and deftly directed militancy. The Bernera Riots of 1872 – as significant an event in the Crofters' War as the Battle of the Braes – brought down an official as powerful as any in Britain. Then there was the shipload of Uig travellers, bound for the Americas, demanding (successfully) that their vessel put in at Port of Ness to put ashore a party suspected of bearing disease. Villains, occasionally; but the people of Lewis are victims never.

In 1914, this was no island cowering before the imminent lash of conscription. Her contribution to the Great War was prodigious; it has been reliably calculated that 6,712 Lewismen served in it, from a population of less than 30,000.

And a very high proportion sailed as members of the Royal Naval Reserve, which launched its Stornoway operations in 1874 – 'in its day the largest single station in the kingdom', gushed William Grant in 1915.

> The physique and efficiency of the Reservemen were surpassed by none, according to official reports . . . This splendid RNR contingent, numbering about 2,000, promptly answered the mobilisation summons of the Admiralty on the memorable 2nd August last year . . . the old Artillery Volunteers (1st Ross-shire), Stornoway, was supplanted some

years ago by the new organisation, the Ross Mountain Battery (TF) and the Stornoway Company thereof is now valorously fighting in the dreary and blood-soaked slopes of the Dardanelles. With the Company are serving 41 of the Secondary pupils of the Nicolson Institute, actually on the working roll of the school when the war broke out . . . Nothing redounds more to the high credit of the Island's endeavour in this world conflict than the voluntary enlistment of young men of Lewis birth or extraction, both at home (though furth of the Island) and in the Colonies, the States, the Argentine and elsewhere in foreign parts. From Canada alone some 250 Lewis lads are now in the trenches in Flanders . . .

Saddler Philip MacLeod, of Steinish, near Stornoway, serving in the Royal Field Artillery, died at the Battle of Mons in August 1914. He was the very first Lewis casualty of the Great War and his name can still be read on the great bronze plaques now affixed to huge boulders just below the stirring Scots-baronial tower of the Lewis War Memorial. This stands proud atop Cnoc nan Uan, a 300-foot summit on the outskirts of Stornoway, and is perhaps the most majestic monument to First World War dead of any community in Britain. And almost 1,000 more Lewismen would fall, disappear or drown before the guns fell silent on the eleventh day of the eleventh month of 1918.

But more names would speedily follow. By September 1924, when a beaten and retreating landlord, Viscount Leverhulme, finally dedicated the War Memorial, it was in honour of 1,150 casualties – a few from subsequent disease and accident, but the mass from the inexplicable waste of the *Iolaire*. That was a final mortality rate of 17 per cent of all who had served, no less than twice the national average.

The loss of HMY *Iolaire* remains the sorest wound. It is an event tinged with corruption – Navy incompetence, political evasion, hard hints of a determined cover-up – and one that must always, sadly, evade definitive explanation. What is left of the *Iolaire* – and it is suspiciously little – decays on the sea bed by the Beasts of Holm. Yet the ship sails on, in abiding island consciousness, as a defining emblem of the protracted woes of Lewis in the years immediately following the First World War. The tragedy served its part, certainly, in the resolve of young men to resist the blandishments of Leverhulme and the wage-slave industrialism he sought to impose. The important land-raids at Gress, for instance, were launched only weeks after the tragedy, in March 1919.

But these hardened young men would besides question the toil and value of an ancient crofting economy and, for the first time, dare to challenge a complacent, controlling, patriarchal order in homes and townships. When they were beaten – as many were – with cold purpose and heavy hearts they left their island. But it proved to the abiding advantage of communities throughout Scotland and all over the world; and, for the most part, it worked out to their own and to the greater prosperity of their children. And they sailed forth on the ghost of the *Iolaire*.

Immediately, the disaster of 1 January 1919 left dozens of families not only deprived of their mainstay in life but, however provided for (scantily by the state and more generously by public appeal), soon finding their relative riches only made them the focus for envy, comment and spite in island villages already sliding into a confounding destitution.

And the *Iolaire*, too – on top of Great War losses and Leverhulme's final failure – not only served its part in triggering large-scale emigration but also denied dozens – if not hundreds – of young women any chance of a partner in life, in an inter-war world where an unmarried woman had neither dignity, status nor significant earning power. And she is but part of years of impacted tragedy so overwhelming that – on an island noted for wit, merriment and song, where even the most Presbyterian women had delighted in tartan and colour and crisp white lace – hundreds were now swaddled for decades in black, fustian and stifling and dehumanising; widow's weeds in which many, many women lived the rest of their very long lives.

In herself, specifically, the *Iolaire* has mocked every New Year since she so desecrated that first one of peace. To this day, for very many people on Lewis, especially those most rooted in faith and sensibility, New Year is a time indeed of quiet joy, but also one of gravity and reflection, as it was always to many we knew in this world and whose loved ones had been ripped from their lives by the Beasts of Holm. 'You go out, my dears, you go out and enjoy yourselves,' the Murray children of Lionel would be assured by their mother each Hogmanay, as she took invariably to her bed, 'but the night has other meanings for me . . . ' Out they duly went, but their pleasure (inevitably) was guilty and thin.

Yet the loss of the *Iolaire* remains, too, a tantalising mystery of the sea – a symbol of bureaucrats in a hurry and of almost incredible maritime ineptitude. She is a ship so elusive we have only one or two certain photographs of her. In a real sense, she is a ship that never was: she was called the *Iolaire* for just a few weeks and never physically bore the name.

The disaster is a story not merely of the worst and most venal in man, but also of real heroism, of determination to survive, of grace given to endure unfathomable emotional pain, of courage to fight for life (and for the lives of others) in an extremity of cold, dark, pain and terror. It is the story of the strength of men who, when all was done, got their heads down and carried on with their lives, through twenty difficult years and yet another grinding war.

The *Iolaire* – insofar as one might try to second-guess Providence – has left, besides, another detail in history.

At the eightieth anniversary of the Armistice in November 1998, the Great War Association – which tends to the needs of veterans – thought there were some 300 still living in Britain. Half a decade later, in June 2003, there were about thirty-two. By 11 November 2008, there were only four of those extraordinary, vital old men. Henry Allingham, 112, is the last to have witnessed the Battle of Jutland, a founding member of what became the Royal Air Force, and the oldest man alive in Europe. Claude Choules, 107, lives in Perth, Australia. He watched the surrender of the Imperial German Navy and delights today over a vast brood of great-grandchildren and their children. William Stone (also 107) did not see active service overseas in the Kaiser's war, but he served nevertheless against Hitler and, enlisted in training by November 1918, is accounted a Great War hero.

And the charming Harry Patch, 110, who has never shed his rich Somerset accent, is the last Tommy – the last to have fought in the Flanders mud. The final Scottish veteran, Alfred Anderson, died in November 2005 – last of the 'Old Contemptibles', the British Expeditionary Force of 1914; and the last to join in the celebrated Christmas truce of that December.

But in November 1998, despite an appeal by the *West Highland Free Press*, no Western Isles veteran could anywhere be identified, and certainly none from the Long Island, despite the depth of its heroism and the scale of its engagement. The last, to my own knowledge, may well have been Murdo MacDonald of Manish in Harris, who died quietly late in 1994, a few weeks after officially opening new Council offices in Tarbert, and who still stolidly attended church at the age of 102. One has to conclude, in all reverence, that one or two who might yet have been with us as the ninetieth anniversary loomed died as a pink youth in the froth and bludgeoning of the *Iolaire*.

I touched the *Iolaire* earlier tonight. There is a certain home not far from where she perished, with sight of Stoneyfield Farm and tucked in the sweep of a bay across from Arnish Light, and here lives a friend who once – distracted and bothered of a Saturday by a tedious neighbour, finally said, half-laughing, half-exasperated, 'Away you go and bring me a piece of the *Iolaire* . . . '

Weeks later, my host came home to find – to his very great surprise – a heap of junk on his doorstep. His friend, a keen diver, had kept his word.

The artefacts are now mounted on a board, and I finger them – a crushed, buckled length of lead piping (perhaps a sanitary fitting, perhaps from steam-plant); a piece of mahogany rail, much gnawed, with corroding hinge and plating; a shell, rather crushed, from one of the requisitioned yacht's guns; two strands of cordite, just two, carefully bonded to the wood; and another little spar of mahogany – all retrieved from the vessel's grave in the summer of 1982.

Yesterday evening, I stood in a Stornoway museum and gazed on her engine-builders' plate, and on her bell. Yesterday afternoon, I was once again on that sward overlooking the Beasts of Holm, the paint if anything more eroded on the inscription. I was struck as usual by the stark beauty of the setting – the sparkling Minch, the hills of Lochs, the sweep of the Broad Bay coast and the humming sprawl of Stornoway, its churches and public places and rows of houses.

Then I headed down, over boggy undulating ground and onto the rock, and still down. You are not far from the start of the footpath before you lose sight of Stornoway, not far from the memorial before you cannot see the lighthouse at Arnish, before you can see no habitation or building or be seen from any. And the rock here is peculiar rock – not the blunt worn gneiss of Lewis, but the extraordinary concrete of the 'Stornoway Beds', a conglomerate thick with protruding pebbles and small boulders, uneven and bluntly jagged, like a huge welded mace.

Even today, bright and cool, it is difficult to keep your footing as you step from one inset stone to another. On the way back up, it is a steepish climb, the ground slick and treacherous.

But imagine it night – black night – and cold; that you have reached this doubtful safety from entire horror, chilled and drained and soaked, badly bruised and bleeding, bare-footed, stumbling into the night and up this brae with raging wind at your back and the lash of spray and the screams of desperate men in the void behind, through bog and tearing gorse in the pitch-darkness – or, yet worse, trapped in a sinking ship, clawing at a buckled and stubborn door, the water rising above your shoulders, the plates looming close over your head as others wrestle and whimper beside you, whimpering to live, the water rising . . .

Or that you are not merely in that darkness but still in this raging, freezing sea, scarcely alive now and barely human, tumbling in the maw of its great breakers, seized and flung almost casually, with the force of a car, against this brutal and craggy rock, again and again and again and again.

Chapter Two

Tolsta 1917

Were You Ever At Garry?

One fine evening a few weeks ago, somewhere in France, an English officer, after taking part in one of the fiercest battles of the war, came up to a North Tolsta soldier and calmly said, 'Were you ever at Garry?' Needless to say, the Lewisman was agreeably surprised to hear his favourite haunts being mentioned on the battlefield. The officer, on being answered in the affirmative, spoke in glowing terms of the beauties of this romantic place and ended up by saying that he thought he could forget even the horror of war, if he could take the wings of an eagle and fly there and be at rest. Let us hope that both the officer and the soldier will be spared to spend a happy evening yet in the vicinity of Loch na Cartach, where one finds in such sweet harmony the beauties of moor, loch, machair and sea.

from the North Tolsta District News
Stornoway Gazette, 27 July 1917

'Gunner Angus MacMillan,' he wrote, in careful longhand, 'has recently visited his parents living at Glen Tolsta. Angus may well be described as a war-scarred veteran. He had seen much service in the South African War. Before the present war broke out he was a sailor, but although he was then fifty years of age he at once rejoined his old regiment and crossed to France with the first Expeditionary Force, and there he has been wounded four times. Out of the Battery with which he crossed in 1914 he can now trace only one man.'

Thus, in elegant, laconic prose, Duncan MacDonald, the village schoolmaster and the Tolsta District News correspondent, on 5 January 1917 first lifted his pen for the brand new local newspaper, the *Stornoway Gazette*. He joined dozens of other modestly paid scribes to maintain assorted village chronicles of births and weddings, retirements and sales of work, Communion seasons and chimney fires, lisping tots on their first day at school and word of deaths in Stornoway, Glasgow, Auckland, Seattle, in a newspage of townships – long dubbed 'From the Butt to Barra' – that still survives in the weekly paper.

North Tolsta lies at the end of a road sweeping from Newmarket, on the outskirts of Stornoway, by the Cockle Ebb and up the western coast of Broad Bay, through some of the best land in Lewis and singularly attractive villages of jolly community spirit, golden sands and tended houses.

She is at once emphatically a place apart – miles over a high and lonely road from the much more pastoral strath of Gress and still more miles on foot by mighty cliffscape from the Butt of Lewis – and a rather beautiful place, her streets and houses rolling down steep braes and up little crags and glens, with an outlook over fabulous beaches and up a rugged coast and across the north Minch to a panorama of distant mainland mountains. Tolsta has a distinct and proud identity, many of the people – blond and blue-eyed – showing their strong Nordic descent. But it was strengthened by considerable nineteenth-century immigration from wider Lewis, and especially from Uig: MacIver remains a dominant local surname.

In the inter-village ribbing that remains part of the Lewis tapestry, the people of North Tolsta are often mocked for a supposed obsessive Sabbatarianism. Certainly the community is finely balanced between the Free Church, who have more communicants, and the Free Presbyterians, who have more people. She is, besides, teased, in a sort of Hebridean urban legend, for an abiding and naturally underground culture of witchcraft. That jibe (originating in a 1930s newspaper joke) has not the least foundation in fact.

Tolsta has long proved a united, close-knit, level-headed place, spared the feudings that have wrecked many an island village and alienated its young. Relations between the two churches, for instance – generally toxic in the wider scheme of the Highlands – are and have always been most amicable. There was a very simple reason for this determined etiquette of courtesy, togetherness and well-tended fences. Tolsta was a fishing community, her men going forth together to win rich harvest of haddock, ling and herring from what was, before the First World War, the very rich ground of Broad Bay.

They fished sustainably – drift-netting for herring, with short-line and long or 'great' lines, each bearing dozens of hooks baited with shellfish. And they watched helplessly through the twentieth century as their waters were robbed by rapacious, illegal and largely unmolested East Coast trawlers, who not only wasted the Broad Bay fishery but regularly swept up and destroyed the laboriously laid fishing-lines of those whose inheritance these waters were. Tolsta lobbied and argued, as government did nothing.

We should not unduly romanticise the arduous lives of Tolsta men. This coast has no natural harbour and, save for one rugged pier, inaccessible to vehicles and at the foot of what, even by Tolsta standards, is a demanding climb, fishing-boats had daily to be launched from the sands and, on return, still more wearyingly, manhandled up and beyond any destructive wave. These were heavy, heavy timber craft, and it took all the strength of their crew – six or eight men – to accomplish this back-breaking manoeuvre. If they survived the perils of so capricious a sea on so exposed a coast, it was this backbreaking chore which finally wore them down.

North-facing and, by Lewis standards, a very high village, most of Tolsta is about 300 feet above sea-level, and some of it loftier. The village is vulnerable to snow, and even today its lonely road-link to the south, if a veritable motorway compared to the tarred cart-track I used enthusiastically to cycle two decades ago, can on occasion be impassable. But there is something most attractive about North Tolsta, from her constant bustle – there always seems to be something going on, from children playing exuberantly, to the trundling tractors, to merry conversation at some house-end – to her many little beauty-spots, from the extraordinary geology below the Cammoch to the sweep of the Tràigh Mhòr, her longest and most majestic beach, and the peaceful idyll of Garry. And there is all the energetic enterprise of a place which, among many other claims to one's interest, boasts the most efficient and least empire-building local history society on the Long Island and has fielded, from a small population, the only village football team ever to win the most prestigious local trophy, the Eilean an Fhraoich Cup.

Nowhere on Lewis gave so much of herself to the cause of freedom as did North Tolsta in the Great War. In the 1911 census, she had a population of 853 – 400 males, 453 females – and by 1915, when William Grant issued his *Loyal Lewis Roll of Honour*, recording every island man at the colours, 189 men of North Tolsta fought for George V – 22 per cent of the village and 47 per cent of the entire male population, including infants, boys and patriarchs. Later in the war – and, indeed, in the rematch a quarter-century later – Tolsta's engagement would be still higher. And, though we might agonise over *per capita* calculations and the wisdom of apportioning grief, HMY *Iolaire* hit Tolsta hard; nowhere else are her casualties more reverently remembered.

The First World War annals maintained by our *Stornoway Gazette* reporter are an affecting narrative of an order long vanished on Lewis. But they afford, too, gritty testimony to past, high infant mortality, the fraught position of widows and orphans, the hazards of the sea, the crushing work of subsistence agriculture, and the menace of infectious disease in a day when doctors were remote, expensive and – in childbed and minor surgery apart – could offer little more, decades before antibiotics, than palliative care and glorified quackery.

It is a tale of wise men, strong women, robust and resourceful children. And all this unfolds against the backdrop of world war, a war to which some sons of the village had rushed home from the ends of the earth to enlist, and the dignity of a people who feared their God, cared for one another, were proud of what they caught and what they grew, proud of their fighting men and proud of the ends for which they served. A place where bright, sophisticated little social events could be thrown, children diverted and little local triumphs recorded against the distant backdrop of U-boats, food rationing, Cambrai and Ypres.

'So well has this district responded to the call of King and Country,' muses our reporter in January 1917, 'that, should the military age be raised to forty-five, not an additional man could be obtained.' But, a fortnight later, he delighted in a soiree of the West Coast Mission, when the 'pupils attending the Bible Class . . . met on the evening . . . for their annual treat of tea, cake, buns etc. They were entertained by Nurse

Stewart, their teacher, assisted by the Misses MacIver, MacLeod and MacArthur, school teachers. Nurse Stewart addressed the children, and thereafter solos were sung and readings and recitations given by the above named ladies. Several hymns were sung by the children. Nurse Stewart closed the meeting with the reading of Scripture and prayer. Mr Duff, the Secretary of the Mission, who takes such a kindly interest in everything that pertains to the welfare of the people of North Tolsta, sent as usual a liberal supply of cake and sweets . . .' None of this sits easily with our tired imaginings of a past, fun-free zone under Calvinist hegemony.

The weather was, by the 26 January, settling after the usual storms of New Year, for 'the two local boats formerly fishing for haddocks have now turned to the herring. They have been pretty successful so far. One of them last week earned over £70 and the other one, although losing half her nets, had a good week also.' There were only two boats now; there were no men home to man others.

But that was a footnote to high drama, which could have had devastating consequences:

Last Saturday evening about seven o'clock, there was an outbreak of fire at 12 North Tolsta. A boy of ten years of age, a son of widow John MacDonald, went to the barn to feed the lambs. Somehow the lambs jumped out of the pen and escaped by a back door. The boy, when he ran out to find them, left his lamp too near the corn. On returning, he found the corn blazing. He tried to put the flames out, with the result that he himself had a very narrow escape. The alarm was at once raised. Fortunately the fishermen were ashore, and most of the men on furlough happened to be living near at hand. Every available man was at once on the scene. As the burning barn was to windward, it was feared that the dwelling house would take fire, and the furniture was removed. At this time the flames had penetrated through the roof of the barn, and the conflagration could be seen several miles away. The situation now caused great anxiety. The houses are so crowded in this part of the village that several men, in order to get at the fire quickly, clambered over the roof of the dwelling-house. Some of the men attended to the neighbouring houses in the case of emergency. Others looked to the thatch of the dwelling house upon which the flames were now freely playing, while most of them, amidst the forking flames, got hold of the couplings and rafters of the burning building, and by wrenching the wood under, the burning roof collapsed with a crash. The men now leaped into the smouldering mass and managed to extinguish the fire. Many of them got their hands burnt and their boots and clothes badly singed. But for the prompt and vigorous action of the men the whole cluster of houses in the near vicinity would have been burnt to the ground. The women did their 'bit' splendidly. Owing to the dry weather and frost, water was difficult to obtain. However, after having exhausted the weekend water supply of their houses, a large contingent of them rushed with their pails to a stream 400 yards distant, and by their keeping up a good supply of water they materially helped to extinguish the fire. The corn, seed and potatoes in the barn were rendered practically useless. Much sympathy is felt for Widow MacDonald, who has in the war lost her husband and her only brother.

By February 1917, others were in sore need of it – Mr and Mrs Donald MacKay, 16 North Tolsta, 'who last week lost their only child after a short illness. What makes the case so very pathetic is that they lost another child last September. The deceased were the grandchildren of the respected missionary of the Free Church of North Tolsta, Mr Angus MacKay, who himself had, not so very long ago, passed through very deep sorrow, having lost his wife and six grown-up children within a few years of one another. Much sympathy is extended to him also who in his own old age has now lost these two grandchildren, who were born and brought up in his own home.'

The haddock fishing continued to prosper – plied by men over fifty, too old to enlist but with the experience and sufficient vigour to harvest the sea. 'Each of the two local boats engaged had one day last week very heavy shots. Single lines in several cases brought in more than two hundred and forty fish each. More luck to them!' But the Great War was not forgotten.

This district has given 200 men to the colours, which is equivalent to 35 per cent of the total population and 50 per cent of the males. There were 124 men mobilised at the beginning of the war; but since then 76 men have enlisted voluntarily. 47 in 1914, 20 in 1915 and 9 since the beginning of 1916. If, however, this district has nobly responded, it has severely suffered in having lost 22 of its bravest sons. These casualties are made up of 7 RNRs, 7 Seaforths, 6 Gordons, 1 Cameron Highlander, and 1 Australian. There are seven widows left with 38 children. Seven of the other men lost have left a large number of total dependents. Including widows and orphans, there are altogether now about 100 dependents left in the district. Widow MacDonald, 12 North Tolsta, has lost her husband and her only brother; Mr Kenneth MacLeod, 56 North Tolsta, has lost two sons; Donald Murray, RNR, 22 North Tolsta, has lost a son and a brother; and Widow J Martin, 76 North Tolsta, and Mr D MacLeod, Hill Street, have lost their only sons. Fifteen of the Tolsta men are interned in Holland, and another local man is a prisoner of war in Germany. The hamlet of Glen Tolsta, with a population of under forty, has given ten men to the forces, seven of whom enlisted between August and September, 1914. An excellent record of voluntary enlistment.

On 9 February, he could delight in another jolly night out, when Miss Donaldina MacArthur was 'on the occasion of her leaving to take up duties in Sandwick Public School, presented with a gold pendant set with pearls and amethysts, as a parting gift from teachers, scholars, and some other friends.' The headmaster spoke and Miss Matilda MacDonald 'one of Miss MacArthur's little scholars . . . fastened the chain round her teacher's neck . . . Miss MacLeod suitably replied for Miss MacArthur. The pendant was supplied by Mr MacGilvray [sic], jeweller, Stornoway.' Island teachers, for the most part unmarried women – this was an age when engagement necessitated resignation – were a resilient lot. Miss MacArthur's quickly appointed successor, Annie MacLean, had come from Stornoway, after teaching in 'Kilirivagh Public School, South Uist [sic] . . . and she had also been teaching in St Kilda for three years.'
And servicemen returned for village respite.

Among those who are presently home on leave are – Sergt. Alexander MacIver, Australian Contingent, son of Mr Alex MacIver, 30 North Tolsta; and Mr Alexander Murray, 29 North Tolsta. Sergt. MacIver emigrated to Australia nine years ago. He spent the first three years there on a sheep station, but afterwards went to Brisbane where he became an engineer. He was very successful in his trade. Two years ago he bought a fruit farm a few miles out of Brisbane. He has now employed a manager to carry on the farm work, while he himself enlisted in March, 1916, and came to this country two months ago. Mr Murray emigrated to Canada ten years ago, and after spending a year or two in the colony proceeded to Chicago, where he qualified in electrical engineering. For some years he was employed as a superintendent with the Canadian Cement Company, and this firm has been engaged for over a year in the making of munitions. Mr Murray enlisted in the Canadian Navy last year. Both are typical examples of our Scottish colonials. While the two were chatting over the reminiscences of their boyhood, a local worthy came up to them and after hailing them in real Highland fashion said – 'What a powerful bugle-call when it has brought the one of you from the Canadian Prairie and the other one from the Australian Bush! Here we have another two examples of the patriotic response made by Lewis.'

So the narrative unfolds. On one February Saturday the two Tolsta boats landed about '3000 haddock between them which was sold locally.' More men came home on leave, one for his own wedding. 'We hope we shall hear of more marriages in the district shortly.' Two more Tolsta lads – 'Evander MacIver, RNR, 11 North Tolsta, and Murdo Smith, Seaforths, 56 North Tolsta' are added to the Roll of Honour. 'The two brothers of the former are already serving – Angus interned in Holland, and John in the Royal Naval Reserve. The father and a brother of the latter are in the Seaforths and another brother is in the Royal Naval Reserve. The two were under 18 years of age. A number of young men from the village working in Rosyth also volunteered several months ago, but as they are engaged in Government work they have not so far been allowed to leave their present occupation.' Again, 'The following seamen have been recently at home for a few days – D Campbell, 54 North Tolsta; Angus MacLeod, 6 North Tolsta; Angus Nicolson, 9 North Tolsta; and William MacLeod, 6 North Tolsta. The last named was rescued from HMS *Triumph* when she was sunk in the Dardanelles. He also took part in the sinking of the ill-fated German raider last year in the North Sea.' And, besides, resourceful Tolsta girls also went south to join thousands in the poisonous air of the munitions industry, their faces and hands soon stained yellow with lyddite.

But spring hastens, and Duncan MacDonald's thoughts turn to food and the perils of crofting. 'During the past three weeks,' we learn on 9 March, 'fully a fourth of the hens in this district have died of a disease recognised by a poultry expert as 'fowl cholera' which he says is comparatively rare in this country though very prevalent on the Continent. This is a great loss to poultry-keepers after having spent so much on feeding of these fowls during the past months.' Before post-war improvement in poultry strains – and the prodigious hybrid-layers of recent decades – hens did not lay through the winter and this was a savage blow for people of most marginal fortunes. Meanwhile, and this would become a heated theme, a large number of Tolsta folk, cottars and squatters without their own holdings were offered potato-plots on Tolsta Farm 'and Mr James Thomson, Tong, was deputed to make the necessary arrangements.' A Board of Agriculture representative duly visited and assurance was given that the Board 'would fence the area under cultivation.'

The constant terror of epidemic could surface. 'Influenza is very prevalent in this district for the past two weeks. A large number of young children are laid up, and there are also many cases even amongst the grown-up people . . . Mr MacIntyre, Stornoway, preached in the Free Presbyterian Church last Sunday, both morning and evening. There were large congregations.'

By month's end there were 202 Tolsta men in the Forces. And our correspondent pauses fondly to obituarise a 74-year-old fisherman – The Bogie 'was often heard to declare that he was never nervous at sea so long as he felt something hard under his feet. No one ever heard him complain of either cold or wet . . . Murdo will be much missed in the district and much sympathy goes out to his widow, his three daughters, and his only son, who is an RNR, and not once at home since he went to the fishing in the summer of 1914.' More men came home on leave – 'Pte. MacDonald is returning at once to France, where he had been gassed more than a year ago, when he was only 17 years of age'; the unsettled weather had disrupted the fishing for the past three weeks, and 'Mr J Campbell, RNR, (54), has six brothers-in-law as well as six brothers on active service . . . eleven out of the twelve had enlisted prior to the Derby Scheme. We think this record would be hard to beat.'

William MacLeod, RNR, 34 North Tolsta, was shortly home with other men in brief respite. 'He was one of the many Lewismen who travelled from afar at the outbreak of hostilities. He came at his own expense from Alaska in the autumn of 1914.' Influenza gained ground. 'This epidemic is very prevalent in the district. The inclemency of the weather is largely accountable for the severity of the trouble. It is said that on account of this illness the school attendance for the past two weeks has been the lowest recorded for several years. Nurse Stewart is kept going night and day.'

Yet the Free Presbyterian Communion goes well and great crowds, too, travel from Tolsta (and through Tolsta, having trekked miles of demanding moor from Ness) to Free Church Communion services at Back. At month's end, 'Corpl. John MacKenzie, Seaforths, 73 North Tolsta, who was a student of Divinity before enlisting, preached in the Free Presbyterian Church last Sabbath, both morning and evening.' There were great hopes for him, if he were spared the shells and bullets, to be ordained to the ministry. 'When young John arrived home from the south after his conversion,' Murdo Campbell records, 'his godly and gifted father would have him engage in prayer at family worship. His parent was anxious to know if the boy had really undergone a saving change, and this he would judge by the nature of his prayer. "Before the prayer ended," as his father said afterwards, "I knew John had grace, but listening to him I was afraid I had none myself . . ." '

1917 was a desperate year on the Western Front – frightful casualties, reverses and uncertainty – and something darkens in the District News. 'Sergt Duncan MacKay, Gordons, 2 North Tolsta, was wounded one dark night four weeks ago by falling into a cellar while he was trying to gather his men, whose billets were being heavily shelled by the enemy. About a fortnight ago his father received a free pass for the purpose of going to see him. It was feared that his case was very serious. We are now pleased to announce that he has so well recovered that he has been removed to a convalescent home. Sergt MacKay was also very badly wounded in September. Indeed his father was then notified of his death.'

But the male of the local species was not so easily finished off, as we read in April:

CLIFF ACCIDENT – A PLUCKY BOY

Alexander Murray, 12 years of age, son of Mr Allan Murray, RNR, 4 North Tolsta, met with a serious accident on Thursday of last week. The previous evening he noticed one of his father's sheep on a ledge on the face of one of the precipitous cliffs on the south side of the Tolsta Ard. His mother and brother tried to dissuade him from going to rescue the sheep, but in the early morning he left home unawares and proceeded to the rocks. He scaled the cliff from below until he reached the ledge, but the sheep tried to clear away, with the result that she brought the boy down the face of the rock – a distance of fully 180 feet. The sheep was killed instantaneously, and evidently the boy was lying unconscious at the foot of the rock for some time. When he regained consciousness he noticed that the tide was coming in, and divesting himself of his boots and some garments he attempted to climb the rock. His mother had been searching for him but failed to trace him. His brother then went off and found him, exhausted and covered with blood, on the top of the cliff. 'Alick,' said he, pointing to the foot of the cliff, 'the sheep is dead and all my bones are broken.' After he was carried home he was attended to by Dr MacKenzie, Stornoway. It was found that no bones were broken, but there are serious internal injuries. It seems a miracle that he was not killed outright. The pluck of the boy is seen in his having when he was so much bruised and battered climbed the cliff, a distance of over one hundred feet. We are pleased to say at the time of writing he is improving.

Even today, in a very different society and when most islanders can drive into town for a Big Shop once a week, spring is a worry for Lewis crofters. Early last century, it was invariably a season of strain, shortage and hardship – stores of peats and potatoes running low, no eggs, scant milk, and no useful grazing. But that of 1917 was unusually fraught in North Tolsta.

'The weather has been very stormy for the past five weeks,' worries MacDonald:

Fishing was practically at a standstill. Last Saturday, however, one of the boats ventured to sea and had a shot of 300 haddocks as well as a few cod . . . Seamen home on leave at present: – Murdo MacLeod (76), John Campbell, junior (54), Murdo MacIver (26), Murdo Smith (24), Alex MacIver (80), John MacIver (69), John MacDonald (23), Allan Murray (4) and Alex Smith (6). This is the first time that the two last named have been at home since the autumn of 1914. Several men are expected home this week, to help with the spring work . . . There has been practically no work done on the crofts so far. But Coinneach Sheòrais and two or three of the older men, who realise the seriousness of the times, have been busily engaged for the past few days delving their new plots on the Tolsta Farm. It is to be hoped the Board of Agriculture will have the fencing material for the plots on the ground as early as possible. This would materially encourage the people to proceed with the work without delay.

He concentrates, naturally, on news – events, details and dramas – but what would startle us if we could walk into that village in this correspondent's day?

We would immediately be struck by the all-pervading peatsmoke from fires that never went out. Then, of course, there would be the housing. A few prosperous souls would, by 1917, enjoy slated, glazed homes of the sort today we know as the archetypal Highland crofthouse – the schoolmaster, merchants, those most fortunate in life or with successful, generous offspring. (There was then a very strong culture, surviving into my own generation, that grown children away in mainland or Stornoway employment sent money home, especially when there were younger siblings to feed.) But most Tolsta people lived in blackhouses and, as recently as 1948, most Lewis homesteads were thatched, a few even enduring, inhabited, into the 1980s.

The blackhouse should not be despised. It had 'seven distinctive features,' as W.H. Murray scrupulously notes,

> four of which are definitive: the walls are drystone, they are double, the roof is thatch, and this roof is set on the inner wall, not the outer . . . The design was evolved to withstand high wind. The double walls could be four to nine feet thick, and the space between was filled with peat, rubble, or earth. The placing of the roof on the inner wall – unique to

the blackhouse – served two purposes; no gale could lift the roof, for the outer wall became a baffle, shooting the blast up clear of the edge, and the tops of the outer walls gave a platform for running repairs on the straw thatch, and for re-thatching every few years, when the men would carefully separate the old outer layer from the soot-blackened inner layer. The latter was set aside for manure, and the former replaced as the new inner layer, on top of which they laid a new thatch with the roots upwards to give anchorage. All was lashed down with ropes of woven heather or straw, weighted along the wall-tops with many boulders . . .

Subsequent close study has found still more sophisticated detail in the Lewis blackhouse. The stones of the inner wall were so placed as to drain moisture into the packed cavity, not into the living space. The cavity itself was filled with a deft mix of boulder-clay, peat and ash, and the best blue clay would top it – the outer dry-stone wall was built slightly lower than the inner, and again water drained off in the right direction, through the grassy sod capping the impermeable clay. The house was built, if at all possible, on a slight incline, for the cattle shared the space with the inhabitants, and the lower, draining, end was the cow's. And the roof, too, was pitched a little lower at the windward end, giving remarkable aerodynamic effect when viewed from the side, and for that very reason.

Yet the Lowlander has long mocked the Hebridean blackhouse. Certainly, they varied widely in quality, depending on the site – cottars and squatters, without title or security to such land as they held, and often harried by neighbours, had little incentive to build the best home. Quality of construction depended besides both on available manpower and the materials to hand. Timber, on a treeless island, was a fabulously precious resource; there are stories of awful drownings as men overreached from the shore to retrieve a floating log. The best blackhouses – good, roofless examples can be seen at Arnol, Bragar and Doune – could be surprisingly sophisticated, with assorted chambers and apartments. The worst – and photographs survive – built by the most hopeless, could be fashioned of nothing more than turf and poor planks; the walls one oozing mass of sod. Such houses were still found on Eriskay in 1935 and, in lean-to form, on South Uist even twenty years later.

But the social order long discouraged the best homebuilding endeavours. Before the Crofters Act of 1886, nobody had security of tenure and there was little incentive to improve a dwelling from which you might be evicted at any time and which, indeed, if done up with any style and skill, would attract the eye of an official, who might then appropriate it for some other tenant to whom he owed a favour. That said, the blackhouse compared well to the poorest urban housing; from what we can read of contemporary slum conditions, the traditional Lewis homestead was infinitely preferable.

The old thatched Hebridean house had virtues that appeal to our ecologically minded age. Thickly built, well insulated, they were most warm, with a central fire that burned day and night and the body heat besides of all inhabitants, be they human, hooved or feathered. The blackhouses were sustainable, built entirely of local materials, thatched with whatever was at hand – in Lewis, usually oat or barley straw; marram grass, heather and even rushes or bracken could be seen in places less favoured.

Assorted busybodies often marvelled how healthy Lewis people were. There were occasional 'epidemics of measles, or whooping-cough, like other parts of the country, and there are sometimes outbreaks of dung-fever – a preventable malady – but altogether the general health is good,' wrote a doctor in 1902. But contaminated water, especially in crowded localities, was an issue. The dung-channel in the byre served as the only lavatory, with men and boys especially encouraged to make shift for themselves among discreet rocks or by the shore, and as Harris Tweed weaving took hold, urine was stored in great barrels for fulling ('waulking') the wool – the stale, ammoniacal fluid efficiently dissolved greasy lanolin. And infant mortality was high. A child had to be tough to survive the first and most vulnerable months, with a form of tetanus (the 'fifth night's sickness', today generally associated with St Kilda, but in truth a problem throughout the Hebrides) particularly feared.

Childbed itself was fraught. Up to the Second World War, a Lewis bride's standard trousseau included not only immaculate layette but her own neatly sewn shroud – as well as a tablecloth for the repast served to the menfolk on their return from the burial; as recently as 1997, we launched a frantic hunt for that 1938 parcel when my grandmother passed away. With skilled native midwifery, and some medical back-up by this stage, complications were seldom fatal. The great fear was 'puerperal fever' – post-partum infection, usually staphylococcus; the first effective antibacterial drugs would not appear till the 1930s, nor real antibiotics until the Second World War. Traditional folk medicine had its eccentricities. It is unlikely that an ailing cow would appreciate being made to swallow a live trout; improbable that the touch of a seventh son cured scrofula, though that superstition of the 'King's Evil', *Tinneas an Rìgh*, endured on Lewis into the 1950s. But others, still abiding on Lewis, are surprisingly effective. Eating two thoroughly hard-boiled eggs will certainly stop diarrhoea; a concoction cooked up from the marsh bogbean is an excellent tonic; a poultice of mouldy bread would, of course, contain crude penicillium bacteria; and an uncle born in 1928, who once tripped and cut his head badly at the Tolsta schoolhouse, still remembers how a capable woman staunched the bleeding – simply by applying handfuls of spider's web from the barn, a natural and speedy coagulant.

This Great War generation, even in peace, lived with death in a way we can barely conceive. Young people died of chest infections; fit men could pass away, screaming in agony, of appendicitis; mothers were lost in childbirth. Practically everyone outlived at least one of their children, and usually several, and the oldest grandchildren besides. Tuberculosis had long snuck into Lewis – there was no history of it, and no acquired immunity – and any outbreak of, say, measles, scarlet fever or whooping cough made parents tremble, and pray.

Now, besides, there was protracted world war.

There was a nerve-shredding theatre to the intimation of the latest village casualty, in both global conflicts of the twentieth century, which no one alive in Lewis in such days can forget. Word came by War Office telegram to the local post office. The postmaster would emerge, sombrely dressed, and make for the home of the local elder (in Lewis, each has a specific pastoral district) and perhaps, if there was one, the minister. The village nurse, too, might be alerted. A little party would then walk out the relevant road to the stricken home. The terror such a procession would cause on that street (where almost every household had a man, or several men, away on active service) can readily be imagined; the nauseating blend of relief and thankfulness and guilt when they approached another door, and how your world lurched when they turned to yours.

Death was not – and, on the Long Island, still is not, though some customs have in recent years been modified – the denied, over-medicalised phenomenon it has elsewhere become. People died almost invariably at home, not in hospital, cared for by their own, laid out for the grave by their own. All windows were blinded until after the burial. All outside work in the vicinity was suspended. Each night family worship was conducted, usually by the elders, at nine p.m., which all relatives and neighbours attended. (In Tolsta, then and to this day, the elders also conducted a morning worship.) The family were never left alone at any time, and relieved of every duty or responsibility, from milking a cow to the preparation of food. After the evening worship, the young people of the village would come in, and converse among themselves – not lugubriously, but without merriment or levity. The grieving slept as they could and, if they could not, there was always a light burning, there was company, there was someone at any hour to fix a cup of tea and a bite to eat.

People regularly viewed the remains, or sat by them. At least two sunrises, and often three (especially if there was an intervening Sabbath, when neither burials nor marriages are conducted on Lewis and Harris), elapsed before the funeral proper. Though deftly incorporated into the new Evangelical order, the origins of this protracted wake are both pagan and essentially practical. The remains were watched (the local phrase for a wake, *taigh-fhaire*, means 'house of watching') until the first slight but irrefutable indications of decay, proof definitive of death. This was no light matter when few doctors ever attended – there is at least one nineteenth-century instance, from Harris, of a woman who revived in her coffin even as the menfolk bore her to Luskentyre cemetery.

At the hour appointed, a minister might preside; more usually, ninety years ago, the elders took that last worship. Then the males of the community would process with the coffin, on foot and in quiet cortege and taking their turn of the load, down the quiet descending road from the village churches to the burial ground of Clach Mhìcheil. A minister, if present, might give a brief Gospel exhortation. Otherwise there would be no further ceremony, save the practicalities of interment, in which all fit men assisted and from which none would turn for home until all was done. Women did not attend, and their presence in the cemetery is still discouraged on Lewis.

Thus, in a most structured, honouring way, the people of the island died and were buried. Part of the awfulness of HMY *Iolaire* in local folklore is how, under the weight of mass bereavement, this system largely collapsed in January 1919, denying many broken-hearted people the normal rituals of grieving.

Could we but find ourselves in this 1917 world, we would be stressed by three realities. Nutritious as it was, the diet of rural Lewis was one of robust monotony, and in spring especially, food could be scanty indeed. We would be disturbed, too, by the extraordinary lack of privacy – almost everyone shared a bed, even grown brothers, or grandmothers with children; people went nowhere alone; nothing, nothing at all could be done in North Tolsta without notice and comment. This was still a world of clan, kinship, communality – one of the deepest love and a powerful moral and social safety-net, an ethic for which two brothers died by the Beasts of Holm; but nevertheless an order we in our aggressively individualistic age would find suffocating.

But most of all we would be aghast at the hard, hard work, domestic and crofting, all the greater during wartime with so few able-bodied men. In just one detail that always makes me shudder, my elderly Shawbost great-grandmother and her youngest daughter had one black spring of the Great War to turn a whole six-acre croft by spade. They had no horse. They had no plough. The two brothers were at sea with the RNR; their father had died in 1912. For weeks they toiled, these two vulnerable women, but had they not so toiled they would not, that winter, have eaten.

We are apt to sentimentalise past, traditional Highland land-use. An exasperated Harris neighbour once growled, in my hearing, of a local holiday-homer, a wealthy English doctor who showed up at his cottage on the shores of Loch Seaforth for a few weeks each summer. He was annoyed that no one grew corn any more, or tended the *feannagan*, the lazy-beds, or kept a house-cow. Why were Highlanders now so lazy?

'That man,' spat my friend, 'would he do what my mother did? I'd like to see him try . . . up and down the brae with a creel of seaweed on her back in the spring, up and down, eighty creels of seaweed a day, weeks on end . . .' It was no easier in Tolsta in the last years of the Great War. Despite the assurances given, the Board of Agriculture did not furnish fencing supplies for the new allotments. 'The Board now say they never promised the fencing; they even refuse to provide the wire netting,' thunders our reporter, 'after the people themselves had undertaken to provide the posts for the fence. However the last word has not been said on the topic.'

There was great frustration, too, over another fatuous failure of authority. 'The people of this district are complaining bitterly of the unreasonable delay in the repairing of the meal mill at Gress. It was announced several weeks ago that the money to put this mill into proper working order had been granted. It is estimated that there is sufficient grain in this township itself to produce at least 200 bolls of meal . . .' It was July before the mill was again in operation, sparing the Tolsta people a fraught journey to the only alternative, at Garrabost on the other side of Broad Bay.

'The fishing has been practically at a standstill for the past three weeks. Bait is difficult to procure' – oil and tar from recent wrecks had contaminated the Lewis beaches – but 'the cattle are all now home from the sheiling and the people are finishing off the carting of the winter fuel. The crops are in excellent condition and it is likely that the barley will be ready for reaping at the end of this month. The potato crop has so far showed no signs of the blight and the plots on the North Tolsta Farm give promise of yielding an abundant supply of potatoes. The hay, however, will this year be much inferior to last year's crop.'

A week later, 'Several young women from this district left for the South last week to start at munition work. This township has sent out a large number of munition workers. Mr MacDonald, 61 North Tolsta, has four of his daughters employed in munition factories. The cattle were sent to the Àrd grazings last week and the boys, while there herding, spent the time in playing shinty. At times there are very lively contests, especially when a team of the South-enders meets a team of North-enders.' The folk of Tolsta play hard: in the autumn of 1916, no less than three Tolsta men were in the cutter that won the Fleet Rowing Race, proudly seizing the Admiral's Cup for HMS *Empress of India*, and there were six other Lewismen in that team, too.

By November, Tolsta could gloat over a record potato-crop. In June 1918, though, we have a flash of the toil involved at the new allotments. 'Last week, the people were busy fencing in this tract of land . . . There seems to be a good deal of healthy rivalry amongst some of the holders – and indeed some of them are worthy of much praise for their work. We know of one man who carried a dozen creels of a special kind of seaweed from Abhainn na Claich, a distance of over three miles. Surely such a man is worthy of getting a good crop.'

Drama continued awhile – this, too, from June 1917:

Last week some children from the north end of the village, after they had got home from school, went out to Glenmore to herd cattle, and as is customary with 'children just let loose from school', they were somewhat playful. One boy, Murdo MacKenzie, 48 North Tolsta, hid another boy's staff in a deep pool of water, closely overhung with heather. After a while he returned for the stick and while trying to reach it he overbalanced and fell headlong into the pool. When his playmates missed him they became anxious and at once began to search for him. After a while one of the boys noticed the little fellow's hand faintly moving amongst the overhanging heather. When they got him to the brink of the pool he was very much exhausted. He had swallowed a great quantity of the muddy waters of the pool before he was rescued. However, after he had been well rubbed by his companions, he was able to walk home with them and although he was confined to bed for a week, he was soon able to move about.

Murdo Murray of 27 North Tolsta was, in September 1917, not so lucky; while bathing off the Tolsta sands, he got into serious difficulties. His frantic friends did their best – one 'managed once or twice to get hold of the drowning boy, but failed to effect a rescue . . . Murdo could not swim and it is believed he lost his balance in the midst of a *cliath* of breakers. In the evening the two boats dragging for the remains found the body some distance from the shore.' The casualties from assorted fronts continued – in one 1917 fortnight, five Tolsta boys were gravely wounded; four of them died soon after.

By year's end, death still more insidious seemed to be stalking the village. By 7 December the closure of Tolsta School has been ordered for three weeks: there is an outbreak of scarlet fever. It will remain shut for two months and it is closed repeatedly throughout 1918 – more scarlet fever, then measles. By late May, two little girls and a young housewife die – 'the deceased's husband, who was at home on leave from Holland, had to leave by the mailboat on Sunday night, just a few hours before his wife passed away. His must have been a case of heart-rending experience.'

Meanwhile, 1918 opened with a terrific snowstorm, the blizzard striking during evening sermon on Sabbath, 6 January. Several lost their way floundering from church; several houses were so buried in drifts 'many families could not get out until they were released by outsiders . . . at the time of writing [25 January] this district is very short of provisions. The mails have been delivered here only once last week. It is feared that much of the potato stock lying in barns has been ruined by the blizzard. It is reported that rabbits have been caught in the doors of the houses in the middle of the village. Although thaw has now set in, it is not likely that the roads to town will be passable for at least a week.'

And over all, as a young lad in Shawbost – Hector MacIver, born in 1910 – would memorialise many years later, there was the pall of war, simply part of the normality of his childhood.

When I look back, I can even recall certain typical Lewis smells: barley seed being boiled in pots on the peat fire or turnips in a bath; the strong acrid smell of ammonia; the smell of manure in the spring, of turnips in the winter; peat being cut on the moor and then dried; sheep being dipped; cows calving; sheep lambing; potatoes slit in raw chips; tea with mint in it; rennet hanging from the rafters, meat being cured in the smoke of the peat fire and it in the middle of the floor of the blackhouse of *taigh Lui*. I remember – though my mother had forbidden swearing in the house – the bagpipe tunes I heard in the village, set to bawdy words, containing obscenities.

After school we sometimes bathed. I had already learned to swim by being flung off the jetty into the sea, by my brothers. But generally, as soon as we were able we had work to do: like herding cattle or rounding up sheep or being butted by horns when putting the ram in; or filling bobbins for our fathers' looms or helping to lay nets or take them in before the Sabbath; or threshing sheaves with a flail on the floor of the barn; or gathering seaweed as manure for the barley field; or carrying water from the well, or planting potatoes if the ground was not too muddy or lifting them; or gathering peats and bringing them home. Because now I began to be entrusted with driving my father's cart. I can see myself as a young schoolboy driving the horse and cart with the red wheels across the moor. My bare feet are dug hard into the blue shafts for balance on the uneven road. I flick the horse gently now and then with a branch of willow.

One word that recurred most frequently in the vocabulary of the entire village was *cogadh*, the Gaelic word for war. And as surely as I had come to accept war as an inevitable and integral part of Lewis life – with its black meal, wartime rations, margarine, contaminated sugar, occasional visits from the Red Cross – I had come to accept the protracted mourning, the three-day wakes upon which the village custom insisted whenever a fresh casualty was announced amongst the menfolk of the village who were mostly in the Navy. The benches and stools that stood around the peat fire were carved with the initials and names of young men on active service, far away from the peat smoke. Any day a set of initials might be written off as the signature of a man who had done his duty, had taken his last wages. When a wake was held for a boy killed in action, I have gone into the house and seen twenty girls dressed in black and the young men also . . .

Hector would once be sent to Stornoway to sell a ill-tempered cow – on his own, and not by road, but straight across Lewis by miles of pathless moor and hill to Stornoway, leaving late at night. He was then twelve years old.

In March 1918, granted again home leave, Corporal John MacKenzie of the Seaforth Highlanders preached in Tolsta's Free Presbyterian Church. 'Corporal MacKenzie had been a Divinity student before enlisting,' Duncan MacDonald again relates, 'and is very popular among the people of the district.' But, before the zenith of summer, as the *Free Presbyterian Magazine* grimly reported,

It is with feelings of the keenest sorrow that we briefly notice this month the death of Corporal John MacKenzie of the 2nd Seaforth Highlanders, which sad event took place in France on Sabbath evening, the 9th June. He was struck by the portion of a shell, with the result that he died of loss of blood as he was being carried to a dressing station. Corporal MacKenzie who, prior to his enlistment, was studying Arts with a view to the Gospel ministry, was a young man of marked piety and great promise, and his death is deeply and justly mourned by a wide circle of friends. He was 'called up' about two years ago, and was for the most part of that time at Cromarty, where he proved himself not only a competent soldier in his country's service, but a useful messenger of Christ in conducting religious services for the spiritual benefit of the men. Before he joined the Army he not only attended to his Arts studies, but regularly supplied congregations on the Sabbath in the Southern Presbytery . . . We extend our deepest sympathy to his aged and worthy father and other relatives in their great bereavement. Their loss is the loss of the whole Church.

He was, lamented the Tolsta District News, 'much loved and highly respected, both as a man and as a Christian . . .' In the trenches, John MacKenzie had organised a little fellowship group with other young believers in the Seaforth Highlanders, and two of them, William Campbell of Arnol and Roderick Finlayson of Lochcarron, became eminent Free Church ministers. Very many years later, as a boy of fourteen, I hunkered by the feet of the aged Professor Finlayson, in an Edinburgh manse in the late autumn of 1980, and asked him diffidently about the Great War. 'Well,' he said softly, gazing aside into the fire and speaking almost to himself, 'I came back, but many others didn't', and, no doubt, John MacKenzie was in his thoughts: a ministry in providence never exercised, a thousand sermons never preached.

In November 1918, Hector MacIver of Shawbost was making for home.

As my friends and I rushed up one afternoon from the hollow in which the village school stood, on the bank of the river, someone suddenly stopped and shouted, 'Look at the flags' . . . and when we asked a passer-by what the flags meant, he said jubilantly, '*Sguir an Cogadh* – the War has stopped.' Not one of us could fully absorb such news; it sounded comparatively meaningless. The world we had known all our lives was a world at war; if the war stopped, it was only common sense that our world would stop too. We were still dazed and uncomprehending when we scaled the brae and halted at the window of the post office to read and transcribe on to our jotters (as was part of our daily routine) the now famous Cease Fire Order issued by Earl Haig. As the white flags waved over the thatch of the houses, we slowly worked out the meaning of the signal, with a pitiful feeling of frustration. As we walked in the street, John MacPhail said, 'They'll all come back home now . . . all the fellows in the war . . . all except your brother, he's killed.' And I suppose it was at that very moment when John MacPhail spoke that it occurred to me seriously, for the first time, that my brother Neil would never return. Until then I had not accepted the fact that he had been killed at the Somme, even though we had a wake for him, two years before. The chattering John MacPhail was the instrument with which I was, for the first time, brought face to face with complete disillusionment . . . Neil had meant to study for the ministry. But the war came and, claiming that he was older than he actually was, he volunteered to join the Seaforth Highlanders. He was too young for enlistment – but not too young to die.

There is no note of rejoicing in the Tolsta District News. On 8 November, influenza is rampant and the school is again closed. 'In a large number of cases whole families are laid up, but medical attendance is difficult to obtain.' A week later, 'There were no fewer than five deaths in this district since last Thursday and four of these are attributed to influenza . . .' The dead include a thirty-four-year-old wife and mother, a seventy-one-year-old matron, a boy of five and twin babies of just five months. 'There is such a deep gloom cast over the district by these deaths that the people feel they cannot rejoice as they should over the glad tidings of peace. One feels like the Gaelic bard after the battle of Waterloo, when he said, "Bha Breatainn dèanamh gàirdeachais. Bha iadsan dèanamh caoidh . . ."'

'Britain is celebrating. They are mourning . . .'

Tolsta had sent 231 sons to the Great War. By the Armistice, forty-one were dead – forty-one men from a village of less than 900 people.

On 25 November, we learn Miss Jessie Fraser, assistant teacher, has just married, and was duly 'presented by the staff and pupils of the school with a silver pot and a cake basket. Miss Fraser was a great favourite among the pupils.' By December the school is still closed and there have been two more losses to influenza – a fourteen-year-old girl and a youth of twenty-one. On 13 December 1918, it is reported that 'John MacIver, RNR, 43 North Tolsta, was a gunner on a sailing ship, when it was wrecked during a storm three weeks ago, on the south-west coast of England. The crew threw themselves into the sea. Although he was crashed several times among rocks, John managed to escape with severe bruising about the legs.' And 'Jessie MacLeod, youngest daughter of Mr Norman MacLeod, 6 North Tolsta, died of pneumonia, following influenza. She was 23 years of age and had always been strong and healthy.'

And then, for four long weeks, there is only stark silence.

Chapter Three

'A Happy New Year, Commander Mason'

In the summer of 1912, ten men from Ness sailed as usual from Port to the distant rock of Sula Sgeir – it can barely be described as an island – for the annual cull of pubescent solan geese or gannets, fat fluffy things on the brink of flight, which (dressed, spatchcocked and pickled) are stored as *guga*, a great island delicacy.

Shortly after they left, a tremendous storm rose, so violent the community feared they could not have survived. And their hopes were indeed dashed when a Royal Navy patrol investigated, duly reporting there was no sign of life and activity on Sula Sgeir.

Some weeks later, my great-grandfather, Angus Thomson of 3 Habost – away on the autumn round of herring fishing – enjoyed refreshment in an Aberdeen pub. He fell into conversation with some East Coast trawlermen and his heart leapt when they mentioned they had recently sailed by this glorified boulder, Sula Sgeir, and how startled they had been to see men there, busy at some weird industry.

My great-grandfather downed his pint, marched forth and telegraphed home. Days later, the merry crew of guga-hunters sailed into Port of Ness, and with a bumper catch. His Majesty's Navy had earnestly investigated North Rona.

The wrong island.

Germany, historian A.J.P. Taylor used to argue, was finally trapped into launching the Great War by her own humourless, micro-managing efficiency – '. . . they had to do so. All the Great Powers had elaborate plans for mobilising their vast armies. Only the Germans merged these into plans for actual war. The others could mobilise and stand still; Germany could not. The German generals and statesmen were prisoners of the railway-timetables which they had worked out in the previous years.'

Just as these Teutonic, inflexible schedules guaranteed war and death for many Germans, so railway-timetables in the dying hours of 1918 spelt drowning for many Lewismen. The authorities had made modestly efficient arrangements to bring hundreds of servicemen to Kyle of Lochalsh by year's end, but given scarcely a moment's thought to how they might then be borne across the Minch.

Of the thousands of Lewismen who served in the First World War, around half – about 3,100 – were members of the Royal Naval Reserve. We have already touched on this organisation. Founded under the Naval Reserve Act of 1859, and a reservoir of manpower for national emergency comparable, in another branch of the forces, to the Territorial Army, it sought men of active nautical experience – fishermen, merchant sailors and so on. From 1862, too, the RNR recruited officers as well as seamen, who wore a distinctive sleeve-stripe of interwoven chain. Drill-ships were established at the principal seaports of the British Isles and reservists, who enjoyed an annual tax-free 'bounty', spent time regularly each year (many compromised on every second year) away at camp, largely for training in gunnery. Men also joined the much larger Royal Naval Volunteer Reserve, for civilian landlubbers, established in 1903. In 1910, the RNR (Trawler Section) was established, with a view to recruiting fishermen and their particular skills, not only in the handling of small craft, but also with nets and tackle – readily adapted for a significant new branch of war at sea: mine-sweeping. In the Great War, casualties among these trawlermen on mine-sweeping operations were considerable.

By 1914, the RNR boasted 30,000 officers and men. A full tenth were Lewismen – an astonishing statistic – and its role in the war effort was distinguished and vital. RNR men secured twelve Victoria Crosses and thought themselves generally (and with some foundation) rather superior to the starchy Royal Navy. It is, though, no slight on their valour, or the many sailors who lost their lives, to note that all British naval losses in the Great War totalled 32,208 – only 4.6 per cent of all British casualties and to be contrasted with the 19,000 British soldiers killed on the *first day* of the Somme.

Many of these men now on their way home had served for the entire war. And there was still just a little uncertainty. We know, of course, the Great War did emphatically end on 11 November 1918. But the Germans had not signed terms of unconditional surrender, merely an Armistice – a ceasefire. It would take the great conference at Versailles to secure a standing peace, which is why not a few engraved memorials of the conflict – those in Tarbert and Stornoway; there are other examples at Great Accrington, Halewood, Swindon village in Gloucestershire, Lurgan in Ireland and many more – quite correctly date the cataclysm as the 'Great War 1914–1919'.

Neither the Government nor the War Office were yet at sufficient ease to order full-scale demobilisation. There were still ongoing theatres of battle, from the mess in Ireland to Britain's increasingly embarrassing part with diehard Tsarists in Russia. Yet martial tension was ebbing. Thousands of English RNR men had already won their traditional Christmas leave. Now, as they entrained back, it was the turn of the Scots, looking to their own celebrations in a land, ninety years ago, where Christmas – not a Presbyterian celebration – was still not a working man's holiday. MacBrayne steamers would sail on Christmas Day into the 1960s, and many Stornoway businesses stayed stubbornly open even until the late 1970s.

Hundreds were making for Stornoway, and special trains had been arranged to carry them from the likes of Devonport and Portsmouth to Inverness, and then to Kyle and its steamer-pier. But what seemed clear, done and dusted to weary War Office clerks, with little grasp of distances or even the geography of northern Scotland and nothing in their heads but train timetables, was about to give Lieutenant-Commander Cuthbert H. Walsh, RNR, the man in charge of 'Movements' at Kyle of Lochalsh, a really bad day at the office.

Walsh was a victim of other men's carelessness. Among his Stornoway colleagues, confusion was even greater. They knew that hundreds of RNR ratings were expected at Kyle on Tuesday, New Year's Eve; but no one had troubled to tell them precisely how many nor even when, precisely, the trains might reach it. Nor could details be soon ascertained. What was obvious to both Commander Walsh and his Stornoway colleague and superior, Rear-Admiral R.F. Boyle, was the near-certain inadequacy of the regular MacBrayne mailboat for the host expected.

The single-screw steamship *Sheila* has had a rather bad press, contemptuously dismissed as 'antiquated' by Nigel Nicolson in his 1960 study of Lord Leverhulme and recalled more with respect than affection by our oldest island residents. She was the hardest-worked and probably the most reliable of the MacBrayne fleet, but indubitably chugged back and forth over the Minch with an almost majestic disregard for passenger comfort.

Built in 1904, she was by no means elderly, especially by MacBrayne standards: it would be 1931 before the venerable Skye mailboat, the paddle-steamer *Glencoe*, finally went to the breakers' yard at the splendid age of eighty-five. She had been built for a new laird of Lewis – Sir James Matheson – in distant 1846, and is still infamous today for a stark sign in her steerage compartment, the only shelter for the poorest travellers: 'This cabin has accommodation for 90 third-class passengers, when not occupied by sheep, cattle, cargo or other encumbrances.'

By comparison, the 1904 *Sheila* was almost luxurious. The *Sheila* was the first MacBrayne steamer with a triple-expansion steam engine, saving much cost in coal; she was a good-looking ship, with 'a grace of outline not so often met with in vessels of her size,' note Duckworth and Langmuir. Under her highly respected master, Skye-born Captain Donald Cameron, she 'fought and defied the Minch, summer and winter, on the passage between Kyle of Lochalsh and Stornoway with passengers, cargo and mails, and the occasions when "weather and circumstances" were so utterly outrageous that she did not venture out were few and far between . . . We have heard of the most fearsome passages undertaken on winter's nights with such wind and seas that, steaming to the utmost limit, practically no headway would be made at all for hours, but sooner or later the steamer would be alongside the pier preparing for her next trip, or return passage as the case might be.'

But she was slow and, at only 280 tons and 150 feet in length, tiny. The night before, Monday, 30th, the *Sheila* had sailed laden for Stornoway, leaving nearly fifty RNR men stranded in Kyle. Even today, there is not much to Kyle of Lochalsh, a rugged promontory by the straits of Kyleakin where, until the extension of the railway from Strome in 1897, there was nothing one might dignify as a village. Commander Walsh had neither the overnight accommodation nor the provisions for hundreds of men. And he may have feared real problems of discipline, given the season of the year; the emotions attendant on the cessation of war; the frustration of men on leave eagerly scenting home, comfort and loved ones; and the normal prejudice of the Southerner against Highlanders. Commander Walsh duly wired his concerns to Stornoway, where they were considered by the local high command.

The Hon. Robert Francis Boyle, MVO, third son of the Earl of Shannon, was a scion of the Irish peerage. Now fifty-five years old, he had taken charge at the Stornoway Depot – the biggest Royal Naval Reserve base in the country – from Rear-Admiral Tupper in April 1916. Boyle took quiet pride in his domain. He oversaw the RNR 'Battery' at Inaclete Point, some admittedly makeshift offices (of necessity shifted from Stornoway Town Hall in March 1918, when the place burned to soot with the loss of the entire local library and the Lewis Estate records) and his own pleasant headquarters at the Imperial Hotel, by the junction of Kenneth Street and South Beach – now occupied by the An Lanntair arts centre.

Rear-Admiral Boyle was a paternalist, kindly to all who knew their place. Like Tupper, he had presented a trophy to Stornoway Golf Club. He was also a sailor by profession and a bureaucrat by temperament. He was vexed at this muddle, the ongoing lack of hard numbers and tedious War Office incompetence, and was determined as few men be stuck in Kyle over the New Year as possible. There was an obvious, expedient solution. Soon after daybreak, Boyle despatched his Stornoway depot-ship, HMY *Iolaire*. There was one detail in his orders that might suggest some lack of confidence in its officers, though there could have been a much more prosaic reason. She was not, he specified, to tie up on return by Stornoway pier, but to drop anchor in the town bay. A Navy drifter would attend and, on successive little runs, land her passengers ashore.

Ludicrous confusion surrounds the most infamous vessel in Western Isles history. Daft as it may sound, the *Iolaire* was not really the *Iolaire*. The ship that died so catastrophically on 1 January 1919 had assumed the name only in October, simply transferred from her Stornoway predecessor to save on the paperwork. Both vessels had, at some point in their complicated history, belonged to Sir Donald Currie – a very grand Member of Parliament, founder of a successful shipping line and a self-conscious laird who owned, for a time, the isles of Scalpay and Pabbay, off the east coast of Skye. Both ships were built as private steam-yachts for gracious living; superficially, they looked very similar. All but one photograph regularly printed of the supposed Lewis death-ship is, in fact, of the other *Iolaire*, which sailed into ripe old age, saw service even in the Second World War and was finally scrapped in 1948.

That ship, the true *Iolaire*, was built in 1902 by William Beardmore and Co., Glasgow, and was spacious, stylish and fast. She had two elegant masts, a steel hull – lighter, stronger than iron and a big selling-point – a 'counter' or sloping stern and a gross tonnage of 999. 236 feet in length and of only 30 feet in beam, she was of racing proportions and quiet, sleek class, and soon a familiar sight in the waters of western Scotland.

In March 1915, amidst what was now a war going badly wrong and an atmosphere of national emergency, it was no surprise she was requisitioned by the Admiralty. Fitted with light guns and other war-time necessities at Portsmouth and painted battleship-grey, the *Iolaire* sailed to Stornoway to serve as official depot-ship at the town's naval base – tendering to larger vessels in the Outer Harbour, dashing back and forth with senior officers, officials and special supplies from the West Highland mainland, and for the most part on glorified milk-runs. In breach of usual Royal Navy custom, the base even took her name. At least one photograph survives on Lewis of a youth – his name now lost – in naval duds and with HMS IOLAIRE on his 'tally' – the band on his cap; but that proves no connection to either yacht, or that he had any part in 1 January 1919; it simply shows he was then stationed at the Stornoway shore base. (An actual crewman would tote HMY IOLAIRE.)

It was in the last weeks of the Great War when this *Iolaire* demitted service as the Stornoway depot-ship, sailing south for other Navy duties in easier times. Her replacement was HMY *Amalthaea*, named for the Greek goddess of bounty, prosperity and plenty and, either by carelessness or for convenience, she simply assumed the name of her predecessor, to whom she bore but cosmetic likeness. She never physically bore it, of course; the names and pennant number of Royal Navy ships were painted out in times of war.

The new *Iolaire* was older, much smaller, and had survived a long career in many guises. Built for Thomas James Waller of 60 Holland Park, Middlesex, in 1881 by Ramage and Ferguson of Leith, launched on 3 August that year as the *Iolanthe*, she was a three-masted schooner-rigged steam-yacht, with counter-stern and of carvel-and-clencher construction. The plaque at Holm today carelessly describes her as 'the wooden vessel HM Yacht Iolaire'. In fact, she was built of iron; that is unambiguously asserted both in the Merchant Navy Register and in the 1913 Lloyd's Register of Yachts. The howling of metal on rock would ring long in the ears of survivors; iron plates and iron ribs still lie on the sea bed.

Iron shipbuilding was already *passé* in 1881. MacBrayne's most famous ship, the gorgeous paddler *Columba*, had been built of steel in 1879, and the steel-built *Windsor Castle* sailed the Clyde as early as 1859 – for, of necessity, iron plating is thicker than steel, a good deal heavier, and rather more brittle, especially in cold conditions. A steel vessel, being lighter, has more speed than an iron one of identical power and dimensions, and might have lasted significantly longer on the rocks: even ten more minutes could have saved many lives.

This *Iolanthe*, the ship that would finally die on the Beasts of Holm, was just short of 190 feet long and of a skinny 27-feet beam. Her capacity was but 634 tons (gross and net tonnage, in shipping terms, are measures of volume, not weight) and she drew a hefty fifteen feet in the water; her steam machinery – driving a single propeller – had a combined might of 110 horsepower. Tracing her career would prove complicated. Mr Waller evidently rose in the world – his latter address is South Villa, Regent's Park, London – and in July 1888, he sold his toy to a new owner in America. Across the Atlantic the *Iolanthe* presumably huffed, but in 1890 she was re-purchased (by Waller, perhaps, though the Register is not specific) and in November 1897 she was re-named the *Mione*; only in October 1899 to become once more the *Iolanthe*.

In 1906, she was bought by Sir Donald Currie, who now dubbed her the *Amalthaea*. And by 1915, when she too was acquired by the Admiralty, her owner was Sir Charles G. Assheton-Smith, Bart., another

London gentleman, who had perforce changed his name from Charles Garden Duff at the behest of a will bequeathing him a considerable estate in Wales. Assheton-Smith – still remembered in posh racing circles – was much more interested in thoroughbreds than boats. He evidently knew his business: between 1893 and 1913, three of his horses won the Grand National.

The *Amalthaea* in 1915 was likewise refashioned as a very light warship – the bowsprit removed, Marconi wireless installed, three light guns affixed. (Even in Hebridean waters, these were far from vainglory. U-boats were frequently sighted, and the MacBrayne steamer *Plover* had a brush with one off Tiree on 29 July 1918, when the Hun attempted to shell her. A doughty Captain Neil MacDougall immediately offloaded his passengers in the boats and prepared to make chase with his own stern-mounted gun, whereon the Germans submerged and fled.) The *Amalthaea* may also have had an enclosed wheelhouse installed – an exposed helm was still common in the 1880s – and a new bridge above it. No doubt carpets and the more fragile touches of luxury were removed; but the fine panelling survived and she had at least three and perhaps four 'saloons', which makes the narrative of the tragedy most confusing. At least one of the saloons was located in the upper deckhouses and others on the main deck below, probably what had been a dining lounge aft and a smoking-room for'ard. She was by no means unsophisticated – she boasted engine-telegraphs and electric light – but she was, nevertheless, inferior in scale and speed to her predecessor at Stornoway.

For, just before the Armistice, she sailed to Lewis and she, too, became HMY *Iolaire*. And the confusion as to which vessel sailed to and from Kyle of Lochalsh that day is not merely a tangle for historians: it may well have contributed to the scale of the disaster.

What of her officers? Her captain, Richard Gordon William Mason, was from Handsworth, near Sheffield, and in December 1918 he was forty-four years old, married to Lucy Lavinia, eighteen years his junior. There were, as yet, no little Masons. His Chief Officer was Lieutenant Leonard Edmund Cotter. He was forty-nine, from Cowes on the Isle of Wight, married to Margaret Eleanor: they had three children. The other officers on board were Sub-Lieutenant Charles Ritchie Rankin, the Chief

Engineer, and his Second, John Hern. In addition to these four gold-braided professionals, there were twenty-three crew. This may sound high. But this was an age of cheap labour, war-time extravagance in manpower, and of much less automated, push-button convenience. Even lowering a gangway was a job for several hands. And – in fact – the *Iolaire* was badly short-crewed this New Year, with about half her complement on leave. The ship's stretched company cannot have appreciated an unscheduled voyage to Kyle of Lochalsh; the lack of deck-officers and look-outs had a significant bearing on events.

Berthing, however, was less than immaculate. Whatever happened, and whoever miscalculated – on the bridge or in the engine room – the *Iolaire* slammed violently into the quay. Too much can be made of this mishap: in March 1966, a highly experienced MacBrayne skipper managed to ground the modern mailboat off Kyle Pier, stranding her long enough, high and dry on a tidal reef, for photographs to appear in rejoicing newspapers. The tides and currents around Kyle have embarrassed many besides Commander Richard Mason, and in most West Highland ports there is no substitute for local knowledge.

The Admiralty yacht sustained no apparent damage, but it is not the impression a skipper likes to make on arrival, and certainly suggests Commander Mason was familiar neither with Kyle of Lochalsh nor the particular handling of this ship. Neither had Commander Mason nor Lieutenant Cotter ever commanded a vessel to Stornoway at night.

Veiled in the Lochalsh twilight, the new *Iolaire* might readily have been mistaken for her swift predecessor. In truth, she was a ponderous Ford Capri to the 1902 Aston Martin. She may have looked like a sprinter: as originally built, she could no doubt fairly shift when under full, unfurled sails. However, as a stripped, steamboat workhorse of the Great War, she was significantly slower than the *Sheila*, and with all the aquadynamics of a bungalow. And the name undoubtedly tricked some passengers. A good number of ratings certainly assumed this was the same large, rather fast and original *Iolaire*, known by repute or from earlier furlough, and – such as had the choice – duly boarded her rather than brave the plodding worthiness of the *Sheila* or, worse still, await the uncertain sailing prospects of the following day.

The *Sheila* was already waiting, on the southern side of the Kyle pier. Commander Walsh's problems had grown. In the early afternoon the latest imperious telegram belatedly advised both that the special train from Inverness was running late – at least two hours late – and (evidence, surely, of a vast multitude) was coming in three parts. Minutes after the *Iolaire* was made fast, there came another telegram, at twenty past four, relating that 530 libertymen had duly entrained for Kyle of Lochalsh at 11:40. There had been not the least urgency to let the local Admiralty know and it was a languor Walsh would not have appreciated.

He moved smartly to retrieve things, instructing one of his adjutants – Lieutenant Hicks – quickly to ascertain just how many men the *Sheila* could take, and made himself for the railway station and the MacBrayne office, as Hicks approached Captain Cameron. Commander Walsh had a simple inquiry of MacBraynes. Was another of the Company's vessels due by Kyle later, one that might be stopped and diverted to Stornoway? He was crisply told there was none.

More frustrated than ever, Walsh stumped back over the pier to discuss the position with the master of the *Iolaire*. A forlorn Hicks was waiting with Commander Mason. The officers of the *Sheila* could not say how many ratings they might accept; they did not yet know how many civilian passengers might alight at Kyle that night seeking a berth on the mailboat. Captain Cameron had already accepted on board twenty-two of the sailors stuck in Kyle on Monday night, and felt he could do no more for the Admiralty just now.

Walsh's jaw tightened. He turned to Commander Mason. 'If necessary, could you take three hundred men?'

'Yes, we could take that easily enough.'

If we are to believe Walsh – remembered, today, only for the worst misjudgement of his career – there was 'a short discussion . . . about boats, belts and weather conditions.' The *Iolaire* had only boats sufficient for 100 men, and perhaps eighty lifejackets and lifebelts; some of the cork-stuffed tabard variety so familiar to us from bad *Titanic* movies, but most a flimsy affair of straps and inflatable rubber. She was simply not equipped for 300 passengers. In truth, Walsh was overwhelmed by administrative incompetence for which he was not responsible. The number of men who finally, that evening, came into Kyle was more than double the figure earlier advised to Stornoway and Rear-Admiral Boyle. Besides, there prevailed a war-time spirit. The Navy no doubt had its own relaxed attitude to the niceties of civilian regulation, and Walsh was eager to put all those island sailors out of his misery – and may, besides, have irrationally feared the consequences of telling hundreds of tired, tough, impatient Hebrideans they would not, after their protracted journey, be home at New Year.

The first part of the train steamed into Kyle at 18:15 hours. The men were quickly mustered; paraded two deep on the platform. The largest, Stornoway, contingent comprised ninety-five files, and they were shortly marched aboard the *Iolaire* by a single gangway; a petty officer at each end keeping count, the master-at-arms checking and stamping the warrants and cards authorising leave. Again the officers checked: yes, 190 men. No names were taken and it is most unlikely anyone thought of drawing up a passenger list. It would be the twenty-first century before Caledonian MacBrayne began to record passenger names on all but the shortest crossings.

There were Skyemen home for leave, too. They were told to stand by for RNR Trawler Division transport: HM Drifter *Jennie Campbell* duly took them, without incident, to Portree. And there were many Harrismen, yet another headache for Commander Walsh. Their orders were immediate and clear: they were to repair to the Red Cross Rest and wait for passage to Tarbert on Thursday, 2 January. The Harris disappointment was most courteously evident.

Now another train huff-puffed into Kyle of Lochalsh. It was 19:00 hours. Walsh watched as the same disciplined military procedure unfolded. This time there were 130 men from Stornoway, and they included Murdo MacLeod, Cèic, of 28 Cross – quiet, with a little beard, and the tired wisdom of a forty-seven-year-old Free Churchman, whose grandparents had been the very last family to live on North Rona.

Walsh and his lieutenants prepared to pipe them all onto the *Iolaire*. At that moment a lad came panting from MacBraynes. Captain Cameron had boarded his civilian passengers; he had done his maths. The *Sheila* could take sixty men. Commander Walsh made an instant decision, and the thirty files on his right of this Lewis contingent were marched onto the *Sheila*. They included all the RNR men of one Lewis village – Tong – and they included Murdo MacLeod, my great-grandfather. The other seventy Lewismen boarded the *Iolaire*. Again the master-at-arms endorsed documents; again, both from deck and pier, careful count was maintained.

There are many stories of last-minute decisions upon which lives probably turned. Two ratings and a soldier – John MacLeod, Norman MacKenzie and a Lieutenant Martin – had been stuck in Kyle all night and furtively boarded the *Iolaire*. After strolling her decks for an hour, they were ordered off by an officer, who told them implacably to board the *Sheila* instead. Later a rumour went about that the *Iolaire* would sail first. One soldier, a Lance-Corporal MacDonald, was scampering across the pier to take advantage when Captain Cameron himself caught him by the arm – 'You come back on my boat, you'll get to Stornoway early enough.' A young Harrisman, Donald MacSween, was pressed by friends to slip onto the *Iolaire*. He was very tempted and could never, afterwards, explain why – just that he had this, you know, *feeling*; but he stayed in Kyle as per orders. All these young men lived.

But 'there was a fellow, Roderick MacKenzie, laying down on the deck of the *Sheila*,' this John MacLeod would recall, 'and he had a bundle of clothes underneath his head. He was asleep. He was a mate in the Trawler section – I think it was in the Trawler section he was – and, after he was asleep for a while, he woke up and he said, "Are we not away yet?" Somebody said, "No", they said, "the *Iolaire* is sailing before the *Sheila*." So he got hold of his bundle of clothes and he went across the rails of the *Sheila*, went across the rails of the *Iolaire*, and that man lost his life . . . lost his life.' And Malcolm Thomson, a twenty-seven-year-old libertyman from Swainbost in Ness, also safely snug on the *Sheila*, saw an old chum from his village, Donald MacDonald, stride onto the *Iolaire*. He had not seen him since the beginning of the war; he, too, jumped one ship for the other. MacKenzie, Thomson and MacDonald would all die.

Commander Walsh had already sent another strained telegram and was now sure this was, after all, the last train from Inverness: come in two parts, not three. It was very dark now. The wind was rising. He made quickly for the quayside by the *Iolaire* for parting courtesies with Commander Mason.

Mason heard Walsh was hovering, and hurried onto the bridge-wing. He had spent a few minutes being avuncular, smiling at everyone, telling one jostle of happy libertymen just to dump their kit in the chartroom – it would stay dry there – urging everyone to make themselves comfortable.

Walsh was rather glad to see the back of the RNR ratings. 'What is the glass doing, Commander?'

'The glass is rising,' said Mason cheerfully. 'Looks like a fine night. We should have a good passage.' Walsh wondered what speed the *Iolaire* might do. 'About ten knots, I reckon.'

'A Happy New Year, Commander Mason – good night,' said Commander Walsh. They saluted; parted.

Walsh hastened back to see off the Skyemen on the *Jennie Campbell*, anticipating comforts thereafter he had thought of all day – a good meal, a blazing fire, his pipe, tea, peace and quiet from those extraordinary Highlanders.

A few minutes later, the pier-hands made ready at the ropes. It was 19:30 hours – half past seven in the evening. From within the *Iolaire*, all along her packed decks, there was great cheer: hubbub of conversation, slapping of backs, assorted Lewis 'coves' hanging over the rail to make jolly, inappropriate and carefully infuriating remarks, and – in Gaelic – some very rude ones.

An emphatic hand from the bridge, an arm outstretched, sweeping down. '*Let go . . .!*'

The hawsers were dropped, snaking over the coping, splashing thickly in the dark waters of the Kyle as the steam-winches aboard ship sucked them in. The bell rang bright and the screws surged.

Shèol an Iolaire, Norman Malcolm Macdonald would write in 1978. 'The *Iolaire* sailed.'

Since March 1973, in a new age that sought to maximise travel by car and with modern West Highland emphasis on the short sea crossing, Stornoway has been served from Ullapool, which – though it boasts no railhead – is readily accessed by excellent roads, with Inverness barely an hour distant. In a time less hurried, when most heavy goods came to Lewis by stately cargo-boat from Glasgow, the long passage to Kyle of Lochalsh and Mallaig was not resented. Were we less obsessed with rapid transit, and the cost of marine fuel not the pressure it has in recent decades become, the Kyle to Stornoway route has much to commend it. For most of it, the ship enjoys the shelter of Skye, and she is not long in the open Minch before reaching the lee of Lewis. The prevailing winds do not attack sideways, beam-on, as they do the Ullapool ferry, but from the stern quarter: on outward passage from Stornoway, the vessel is largely 'stemming' them.

The journey, nevertheless, calls for responsible seamanship and an alert, experienced master. There are strong tides, and to a sailor the tide is not a vertical force that lifts and lowers your dinghy at the jetty, but a current which can be as determined as that of a full river. Stornoway Harbour itself is prone, in unsettled weather, to 'seiches': most rapid changes in local sea-level – some two feet, either way, in ten minutes. Early on passage from Kyle, too, the local tides are anomalous, with the flood-tide flowing not north (as is the norm of the west coast) but south down the Narrows of Raasay, with interesting consequences in certain conditions.

And the Minch is a notorious piece of water, as challenging and capricious as – and in some respects much more dangerous than – the open Atlantic. She is remarkably shallow in spots, her bed ridged, mountainous, plunging; that impacts considerably on the movement of the sea. Even today, with all the aids of navigation – radar, sonar, the sat-nav which has now displaced post-war Decca – the Minch demands profound respect. Ninety years ago, safe and successful passage at night depended on the good old skills of chart-reading, dead reckoning, and of knowing where precisely you were, where you were heading and on what bearing and at what speed, throwing tide, current and weather into this constant calculation. Your only aids were charts, compasses and a handful of major lighthouses – the light at South Rona, to port, atop the straggle of Raasay; and the one at Rubh' Re, north of Melvaig on the Wester Ross coast and by the mouth of Loch Ewe. These were important. But two were critical, that night, for HMY *Iolaire*: Milaid Light on Kebbock Head (A' Chàbag) on the South Lochs coast of Lewis, and Arnish Light by the welcome entrance to Stornoway harbour – and there was a third, too, both as aid and as complication, the lighthouse of Tiumpan Head at the last jut of the Eye Peninusula, well north of Stornoway and the most easterly point on Lewis. And you are well clear of South Rona, and of Skye, before you can see either Arnish or Tiumpan.

Had this been a calm night, a clear one, the skies cloudless and starlit or, better still, with vast and kindly Moon, our tale would end very differently. This was not such a night. Conditions were by no means dangerous for a ship in the right place, but they grew progressively unpleasant, a night when men would not choose to be out at sea save in the execution of duty, and when those who were might deem it prudent to make for port. And a ship aground, on rocks exposed to a building wind, would be in mortal trouble.

By 21:55 hours the *Iolaire* had evidently cleared South Rona and made an alteration of course, as normal, until she was south-east of Kebbock Head and the Milaid Light, when her helm would swing again, as normal, to bear on the Arnish Light. Though increasingly choppy, the night had been clear. From midnight the wind would markedly freshen from the south and now she waded through squalls, with light but skirling rain and, as she chugged on, perhaps lashes of sleet.

In the sentimentality of the times, the scenes aboard the yacht were, afterwards, heavily embroidered; partly to over-egg the pathos of it all and partly, perhaps, as some sort of awful warning. We are still invited to imagine joy, good cheer and bonhomie, men relaxed and happy, greeting others they had not seen for years, laying amused eyes on fellow RNR veterans they had last glimpsed at the peats of 1914 as downy young boys. There would be music – surely there would be music – songs sung, and stories told, and the New Year itself finally hailed with exuberance, as in warmth and comfort and in full ceilidh-party unawareness they proceeded to their doom.

This is not generally sustained by the hard evidence of survivors and makes little sense when the facts are considered. For one, these men were – for the most part – utterly exhausted, having cooled their heels on chilly Kyle pier with no opportunity of rest or refreshment after long, long journeys by spartan troop-train. For another, there was precious little comfort to be had. The *Iolaire*, the *Amalthaea*, had never been designed as a wholesale passenger-carrying vessel, far less one for overnight winter slog in chill northern waters. She had been a floating summer-butterfly for one rich family, and would have struggled with even 100 servicemen.

Tonight, she was grossly overloaded. Only the first and most fortunate had secured seats. Many could not get under cover and some struggled even to find sufficient open deck where they could tug off their boots, lie down, and nap as best they could. Some survivors attested to jamming the chart-room and even the galley.

Of course there were reunions, there would have been much initial bonhomie – charged, not least, by getting clear passage to Lewis at all – and, no doubt, pipes were lit, and the youngest fumbled for their cigarettes, and such food and refreshment as men had (for there was little available on board) brought forth to be cannily shared or covertly consumed. At least one melodeon was played; there was some singing and, come midnight, warm exchanges of New Year felicitation. But the sea was no novelty to these men. And they were so tired. Many soon dozed. Others talked on quietly, barely overheard, voices coming in close from far away, in that faintly nightmarish manner we remember from early childhood, as boring grown-up people spoke of boring grown-up things late at night near the end of an epic drive, as you squirmed half-awake in the back seat of the family car.

Each sailor tonight had his thoughts, his anxieties, his expectations for the future, and the immediate and wearying worry of the last journey ahead – few lived in Stornoway and most had miles further to go, with little hope of even horse-drawn transport, though there was quiet speculation of hiring possibilities: a boat, even a car.

There were grey-headed fathers – the oldest man aboard, Seaman Angus MacLeod from 11 South Shawbost, was fifty-one – and a good few had served since 1914 (though, contrary to mythology, all would have enjoyed occasional home leave). But there were very young men, too – lads in their teens – who had been no further than their basic training in the south of England. The youngest of all, Signalboy David MacDonald, lowest in the hierarchy of the *Iolaire* crew, was a pink Aberdonian and just seventeen.

No longer, in any meaningful sense, under Navy discipline, and free to dispose of themselves as best they could in packed space, men naturally sought the company of fellow-villagers, which may have had significant bearing on the final casualty figures for individual townships. But every man there had his own tale. Recalling the night very many years later, Donald Murray of Tolsta could still taste the delight in reaching Kyle – 'many of the village lads were there; many who had not met since the start of the War. There was real fellowship . . .' Dòmhnall Brus, though still only twenty-three, was a tired, battle-hardened old warhorse. He was glad to be going home.

Seaman Kenneth Smith, forty-four, from Earshader – in Uig, by the narrows of Bernera – had had time, on his travels, to buy gifts. He had shrewdly grasped what would enchant the womenfolk, and his kitbag was replete with fine things, including a silk scarf and a shawl no less lovely; these he had pounced on in Gibraltar, for his young wife. He, too, had served through the whole Great War – from the docks of Chatham to the decks of HMS *Cove*, and he had spent most of it splicing rope (there were three Lewismen in that squad, greying as himself, one from Bernera and one from Uig and another from Point). They were proud of what they did, a skill done at its best by few indeed, unwrapping and welding hempen rope together in ingenious joins that could bear a vessel under tow, proud of four years at that ceaseless toil as a 'bosun sparkie'.

To his four-year-old daughter, his Mòr, or Marion as the forces of English and power would have it, he was an entire stranger – she had seen him but once on leave, and she thought her *Seanair* was her Dad. No wonder – with his sisters about her and genuflecting to the patriarch at every opportunity, *a Dhadaidh* this and *a Dhadaidh* that . . . well, it had vexed Kenneth a bit in September when he had got leave for his son's baptism, but home was just hours away. The Lord willing, he would spend the rest of his life at home, as Abraham, sitting in the door of his tent, and he would go to war no more.

His lady would light up when she saw the silk. Then she would pretend it was not of much account, and look at him sideways, before the mask slipped, and later – about the chores and toils of a wife of the sort he had scarcely dared to hope for, she would be singing through her duties, the songs of Shawbost and of Uig, the psalms of David and Asaph. And he thought, besides, like any sensible man, of dykes and cows and peats, and if they had remembered to order the molasses, and what mail might await for him. And then Kenneth Smith nodded off for a wee *norrag*, for the day had been long and he felt every hour of his forty-four years.

Seaman Kenneth MacPhail came from Arnol, on the West Side. He was twenty-eight years old and had already stared down death in an extraordinary deliverance. In October 1917, his ship had been torpedoed off the North African coast; he was the only survivor and crawled at last onto dry land only after thirty-six hours floating in the warm desolation of the Mediterranean Sea. Now he was going home, fatigued, thankful.

Donald Morrison, Am Patch, from Knockaird in Ness, of fine bearing and lugubrious jaw, was – at twenty – one of the babies on board. He was of hardy people: his splendid mother, Jessie, would still be driving cows over the local moor in her mid nineties. Called up in March 1917, the Patch had served in the English Channel and finally made it into Trawler Division as a certified gunner and gun-layer – a bare two months before the end of the war. He had already been granted four days of compassionate leave in December when his father took gravely ill; but, to his anguish, the Patch had been forced to return to duty at Portland even as the *bodach* failed. Word soon reached Portland, of course, that his father had died.

So it was a heavy-hearted journey for Am Patch – a bereaved home ahead of him, a grieving family – and Donald had struggled to hide his emotion when, alighting at Kyle, he saw his brother Angus, fourteen years his senior, waiting for him. Angus Morrison had reached the previous day, but had declined evening passage on the *Sheila*, so he could surprise young Donald and go the last miles together. And perhaps, besides, he would find it easier facing his widowed mother, on so poignant a New Year, if the wee fellow were there, too.

John MacDonald was another *Niseach*, thirty-two, from Skigersta. He had just got married, winning a brief leave to say 'I do' to his

Jessie on 18 October, back home. They were very much in love. Alex John Campbell, from Habost, had made it aboard the *Iolaire*, but was distinctly cross because he had not been able to win a berth for his brother John, still on Civvy Street and returning from hospital treatment. They had argued and cajoled, but to no avail. John had not even made the *Sheila* and was stuck in Kyle for the night.

John Finlay MacLeod, Iain Mhurdo, 4 Port of Ness, was a boat-builder. Well, that was what his people did: they built boats, most broad of beam, with curiously sheered prow, and clinker-built timbers, and oars that thin the ignorant sometimes teased they were actually those chop-sticks that a 'Chinee' would be at his potatoes with. But this sort of boat, the *sgoth*, had been first engineered by his distant Viking forebears and was perfectly suited to the waters about the Butt of Lewis.

Being a boat-builder was good. Being the baby of the family, the *biodan*, was bad. 'Am Bìodan,' they had crowed at times in the Lionel School playground, when his temper had flared in the youngest son's constant, desperate hunger to be respected, 'Am Bìodan . . . O, a ghràidh! O, m' eudail bheag, bhòidheach!' and he had, no doubt, blushed and seethed, the little John Finlay, but would not weep – no, he would not show them he hated being a *biodan*, with all its connotations of sweet and helpless and Mammy's wee darling.

He knew his boats, John Finlay MacLeod: more, he knew the sea on which they must fly. He was not quite yet of the calibre of such Ness seamen as the patriarch in Lionel who could walk out on Tràigh Shannda, lie down flat out on his stomach, put an ear to the sands, and then announce – with unfailing accuracy – where the storm was and when it might come.

Nor was he quite of the authority of that old, weathered *Niseach* skipper who had suddenly turned whey-faced, after a glance at the horizon and another at the scuppers, and ordered them to turn at once and make for port. He was not one to be trifled with. They had belayed all and turned and rowed – rowed and rowed – and he was at them, the old man, the *bodach* and the *sgiobair*, to go faster, faster: a lad had demurred, and next they knew the knife was out. Like any self-respecting Nessman, it was that honed it could shave hairs from the back of your hand. He told the boy – he told them all – he would stick this in the guts of the next man who stopped rowing. They did not argue. They did not even think of arguing when, nearing the surf, they realised a dreadful squall was hard on their stern.

When they grounded, exhausted in the sands, the boat fell apart about them. In the shock of that moment, they splashed in consternation the three or four paces to dry land. The skipper was at once relaxed, a little embarrassed, rather triumphant. He explained the keel had split, a mile and more from shore – he had seen it crack, right across – and known, with terrible certainty, the moment they stopped rowing and let her flex and loose and sag, the moment that forward impetus was gone, they were all dead men.

John Finlay had not quite the experience to match that. Not yet. But many a day he had watched the breakers and the surf on the sands of Port of Ness, and he had learned much about waves, and the patterns of waves, and the third big wave . . . Some day, he, too, would have respect.

John Murray, Iain Help, had been born and brought up in South Dell. He was thirty-one, and no one could accuse his family of faltering in their duty to George V: all six brothers were on active service. The authorities – as was the custom, if only to ensure the parents were guaranteed at least one surviving support in their old age – had offered Mr Murray, back in South Dell, the opportunity to recall a son home. But which one? An earnest Christian, the crofter prayed fervently and long – and decided, as they will still tell you in Ness today, it was the Lord's mind that all his sons should continue in the First World War. His descendants in 2008 still have a splendid letter – dated 20 October 1914 – from King George V himself, conveying his gratitude for Mr Murray's sons, and His Majesty's 'best wishes for their success, health and happiness in the noble career they have chosen.'

Kenneth Campbell, 54 North Tolsta, could go one better than that; his mother, facing the same choice, could not make it either, and – leaving it in the hands of the Most High – had let her seven boys go to war. Kenneth was twenty-nine. He was not long married and young enough to look forward to the pleasures and prerogatives of the wedded state. He had posed early in the Great War for a photograph with his brother

Angus, and they had folded their arms and stuck lit fags in their mouths and tried to look, as we might say today, well hard. But at the critical moment that the shutter fell, Kenneth had been looking away and his brother, besides, had made him giggle. Still, there would be many more years for photographs.

John MacDonald, 25 Lower Shader, was thirty-two, manly, confident, and unmarried. Hopes boiled with anxiety. He had someone in his mind and on his heart, and in his pocket he had, tiny and carefully wrapped, the surprise of her life.

There was another group, too, who attracted glances and some wry comment: these travellers were betrayed by their darker appearance, their coy separation from everyone else, and a Gaelic and an accent – lilting, liquid, aspirated, with sliding fluting vowels – quite unmistakable. Despite Commander Walsh's explicit orders, eleven Harrismen had boarded the *Iolaire*. This was precious leave and they had already lost much time, only to learn on Kyle pier they were expected to stew in that Nowheresville for two nights more. It is understandable if they decided to 'mishear' their instructions and, as hardy as everyone else, they were quite ready for the long march on foot over moor and mountain to their own parish. It would be some trek – especially for the Berneray lads – but Tarbert, Scalpay and the Bays could be reached within the day, if the snow held off, taking the hill-pass by Maraig to Urgha. The Harrismen would certainly not linger in Stornoway, a town some of them had never even seen.

Deckhand Finlay Morrison, 25, from the Isle of Scalpay, headed homewards with mixed feelings. Had there been an understanding with Ciorstag Ruairidh, or had there not? A promise, at least, to be engaged at some convenient season in the near future? Being fair, nothing had ever been said. But she had not, after all, waited for him, and with the uncertainties of war and another fellow at home tipping his cap in her direction, and the tacit longing of most young women to be mistress of their own hearth, to be secure with their own provider – yes, he could see and almost accept her logic. He might look just a little cross-eyed at her smug husband. Still, there were other girls . . .

Robert MacKinnon was a sturdy fellow from the Caw, a tiny

hamlet peeping over Tarbert from a precipitous brae. And there was Alexander Campbell, from Plocrapool – everyone called him Alick, though – and he smiles shyly at us from his RNR photograph, still juvenile, pretty-boy. And there were the Berneray men, and MacLean and MacCuish from Northton, and Morrison from Stockinish, and John MacKinnon from Tarbert . . . but the *Hearaich* were as tired as everyone else, and uneasily aware that their presence aboard was not, perhaps, generally appreciated.

Not a few Tolsta men had won their way to one pleasant saloon, negotiated its narrow door by the press of bodies. There and ever so quietly they crowed, but as night lengthened and minds dulled they huggled into their sea-coats, leaning on one another's shoulders, heads stretched out on convenient laps. They slept as best they could like a basket of puppies.

Men from Park and men from Point, men from Uig and men from Carloway; the Bragar men and the Broker men; Cromore men and Crossbost men; the coves of Stornoway and the lads of Lochs . . .

The *Iolaire* hastened onwards with her brood. There were men from Point, of course – many men from Point, one of the most crowded Lewis districts, with emotional people of radical outlook. Alasdair MacKenzie, Am Boicean, of 1 Aird, was delighted to collide with his cousin, John MacKenzie, Iain Iain, of 5 Portvoller. They had not met since the outbreak of war. They found a corner of floor, eased themselves down, did much catching-up. Someone started playing a 'box' in this packed space – old tunes, Lewis tunes. It turned out to be still another cousin, Alick MacKenzie, Dài, also from Aird. Then Alick MacLeod of Portnaguran joined them, beaming. And all over HMY *Iolaire*, long-lost friends and seldom-sighted kin were thus rejoined.

Squashed or chilly as the RNR men were, they had at least entire relief from any immediate responsibilities. Up on the bridge, Lieutenant Cotter had now charge of the ship – Commander Mason had retired to his quarters at 01:00 – and was, if the men below had but known it, rather floundering. This wretched shortage of crew – no one with him but the helmsman, and that but an obedient voice repeating orders through a tinny pipe; no one to spare as a look-out, and the weather slowly, definitely deteriorating, the wind still rising and the seas with it, and the rain and sleet coming in vexing curtains. Leonard Cotter checked charts, bearings, and steered on to the distant lighthouse, one wet, winking glow in the black velvet before him.

'At first the night was not too bad,' Donald Murray would recall, 'but it was bad in the Kebbock Head area and to the south – it was bad there. Anyway, we did not doubt that everything was fine, but I remember very well when we were opposite Loch Grimshader, saying to Dòmhnall Red, "I am going to run up to see how far we have got," and I went up to the deck and the light of the Stornoway lighthouse was almost flashing in my face, almost directly opposite Grimshader. You would think that nothing could go wrong, but now there was a strong breeze on the wind and it was behind us and when it struck the vessel the sea was fearsome. So I went down. "Well, lads," said I, "she is almost at the lighthouse . . ."'

In fact, had he known that Mason had just vacated the bridge for a quiet kip, Donald Murray would have been immediately troubled. A ship's master frequently leaves the bridge during passage – may quite legitimately take some shut-eye – but he remains nevertheless fully responsible for his vessel and, when approaching harbour, his place is emphatically on the bridge.

They did not dream of challenging officers, of course, but some – fishermen, especially, who knew the Lewis coastline and the Stornoway approaches intimately – grew uneasy as they squinted through the murk of the night. They could not quite see where they were, but they smelled something untoward; something of land that was too near, and lights in bearings that did not make sense.

Still, going home . . .

'Every one of us had a kit-bag and we hoisted them up and we were just going on deck as she came under the Light. After coming on deck we were aware that she had altered course to the East of the entrance instead of going in . . . agus chan eil mi a' tuigsinn; I do not understand! . . .' and there, on dated, prosaic cassette-tape, Dòmhnall Brus is reduced, for long moments, to weepy incoherence.

That inquiring parade of the decks almost certainly saved his life; and the detail he recalls may be of critical importance. But Murray was not the only one looking out, puzzled, pondering. Seaman John Montgomery, of Ranish, could make no sense of the course the *Iolaire* had chosen either. 'I could see the sea breaking on the shore . . . The land I saw was the land on the East side of Holm Bay.' Those already up, alert and making ready with their gear, were intrigued. Someone wondered if this was a special new route the Admiralty had developed: a smart-alick shortcut. 'Nach e tha dol faisg leatha!' admired a man from Lochs. 'Isn't he going close!'

The Patch beamed at a Eorodale buddy. 'We'll go on the rocks here, Calum!' 'Aye,' cracked Malcolm MacDonald back. 'You *and* those Harris boys . . .'

Stornoway is one of the finest harbours on the west coast of Scotland – the outer deep and fairly sheltered, the inner curling in on the town like a winkle, as James Shaw Grant once put it, affording absolute haven in all conditions. But the harbour entrance is only 700 yards across and the marks and lights have not significantly changed since 1919. There have been, however, two important reforms: the erection of a light on the eastern corner of Goat Island, and the 'zoning' of lights by colour: if you approach in the white zone of Arnish Point light, for instance, you are safely in channel: in the green, you are in the wrong lane; and in the red you are heading for shallows, rocks and land in a way that is wet, expensive and dangerous.

Both actively encourage vessels to hang close by the western side of the channel on entry – Arnish Lighthouse (marking the headland) and Arnish Light Beacon (on an extending reef) glow on the same side – and there are excellent reasons for that: several treacherous areas of shoal water, especially that drying skerry north-east of Arnish Point and another from the north shore, and the Beasts of Holm – 'rocks a cable SSE of Holm Point at the east side of the entrance,' advises a 1996 edition of *The Yachtsman's Pilot to the Western Isles*, 'which dry up to 2.3 metres and are marked by a green beacon 5 metres high.' It is not, though, lit, any more than there was a light on the Beasts on 1 January 1919, though a skipper today bearing on this nasty, lurking reef would find himself in the warning red zone of Arnish Point light.

Montgomery marvelled at the wash on the Holm shore, and noted he could no longer see the light at Tiumpan. The flash of Arnish was straight ahead. Down in the saloons, lads had congealed all night in dozy ceilidh. In the smoking-room, there talked a little knot from Ness, including Angus and Donald Morrison of Knockaird, John Finlay MacLeod of Port, Malcolm MacDonald of Eorodale. Of practical mind, they were planning the last leg of travel. Might it not make sense to find Stornoway digs for the night, and buy passage come daybreak on a drifter or patrol boat that could drop them off at Port?

A crewman bustled by, trying to look important. 'Get your gear ready, boys,' grunted James MacLean, a Campbeltown fisherman. 'She's just coming into harbour.'

On the open starboard deck, Montgomery's concern was spreading. Those about him could not believe their eyes. Men started to murmur. One yelped, 'He'll run us aground . . . !'

The *Iolaire* turned, turned to port, rather urgently, and something changed markedly in her motion.

Angus Morrison stretched, stood. He had to hurry down the ship to retrieve a parcel. He pivoted, barked at Donald. 'If you get lodgings for the night first, be sure to keep a bed for me!' 'Gu cinnteach,' said Am Patch, 'surely . . .'.

His brother was several strides away when she struck.

Chapter Four

The Beasts of Holm

There is no need to labour the discrepancies . . . until disaster struck: there is both too much and too little information for us to be able to resolve them . . . it is not altogether clear what happened . . . Eyewitnesses are renowned for their unreliability; when there are so many, their numbers increase rather than reduce the inconsistencies so that it only remains possible to approximate. The prospect of death may well concentrate the mind; but not, it seems, on fact.

Robin Gardiner and Dan van der Vat,
The Riddle of the Titanic (1995)

John Finlay MacLeod had just emerged from the smoking-room, was closing the door – the knob was in his hand – and there was a bang and a screaming of metal on rock, and over he went, on his two hands. The man who had been ahead of him staggered, turned. 'What's that?' 'I don't know . . .'

The collision threw many off their feet, into one another, slamming into bulkheads and fittings and railings, tumbling sideways as the stricken yacht piled her bows high on the reef and, within half a minute, keeled to starboard, her gunwhales dipping into the froth.

Such was the impact that Donald Morrison – who, after all, had spent weeks chasing the wretched things – immediately thought, 'We've hit a mine . . .' He remembered, always, how the ship shuddered and stopped, and then rolled, in a slow but sickening movement, on her side. 'A lot of men jumped off, I suppose fearing that she would overturn . . .'

The *Iolaire* did not capsize, but within seconds of impact dozens were helpless in the sea, in a maelstrom of rocks and brine and her own crushing bulk; some may well have jumped, but most were flung over by the smash itself, or washed overboard before they had even time to regain their feet, and none of those men lived.

The great billows from the south had now open season on the *Iolaire*, a ship as doomed from the moment of her strike as if she were ablaze from stem to stern. It was the middle of the night – pitch-black night – and what had, seconds before, been uncomfortable conditions for a moving vessel became the hounds of death on one stranded. There were shouts, roaring and much confusion; voices of alarm as, most quickly, these seasoned seamen grasped the gravity of things, even as waves exploded over their ship and the screams of the drowning rang round.

Only a few had yet any clear idea where they were. But the *Iolaire* had wrecked herself right inside the Beasts of Holm and in the worst possible conditions – 'caught between the wind and the sea in total darkness on the one hand,' writes Tormod Calum Dòmhnallach, 'and, on the other, a steep and rocky shore constantly pounded by huge waves.' Worse, though so near port – only a few hundred yards from the harbour entrance and half a mile from the quayside – she was quite out of sight of town, of help. Even had it been daylight, the Beasts of Holm lie in a singularly lonely spot, with no house or community in view, and the adjacent shoreline is still difficult to reach quickly. Only the highest mastlight could be seen from harbour, peeping over Holm Point, and it afforded no sense of the life-and-death struggle below.

From Stornoway in 1919, with so little motorised transport and the metalled road stopping much shorter of the locality than it does now, the chances of any aid arriving in time were remote. To cap everything, she was not only on the Beasts but behind the big outer rocks, cornered on the landward side. There was no hope of help from sea, for no vessel could reach her and herself avoid certain disaster.

Thus the *Iolaire* lay, holed and helpless, plates shrieking and tearing, exposed without mercy to the one wind from which she had no shelter, a southerly – perhaps Force 5 or 6, and building – howling – up the Minch and off the Lochs coast over a considerable 'fetch' of water, powering enormous combers to hammer on the casualty. Had it been flat calm, most would have survived. Had the ship even been held absolutely fast, utterly stranded and locked onto the reef, that would at least have bought time. But, while captive, she was not held fast, her very hull now a handle by which wind and wave could wrench and twist the vessel about the rocks all the time, with constant damage as, remorselessly, she flooded. To cap everything, the tide was on the surge, only increasing the forces assaulting HMY *Iolaire*, denying further minutes on which lives depended.

Yet her lights still blazed, her engines surged on chaotically for at least half a minute, perhaps more; and the telegraph rang, the bell echoing through decades in the memories of such as would live, amidst cacophony of wind and sea, alarm and rock and iron.

The *Iolaire* was done. What follows is inevitably a chaotic narrative, or many little narratives, each man with his own story and none with a clear overview of the whole, but seeing what he survived and remembering what he could bear, defined by that experience for the rest of his life. Some proved heroes; others moved with thoughts only of saving themselves, and others waited for orders, not sure what to do, and thus unmanned were overwhelmed and drowned.

But the central things of the disaster were all in William Grant's report for the *Stornoway Gazette* of Friday, 10 January 1919, and – despite the pressure of time and the most limited resources – not one, then or ninety years on, has ever been gainsaid.

There was a crash. The ship heeled over. Fifty or sixty men fell into the sea and were all lost. No clear orders were heard from the bridge. There was much confusion, though no panic. It was pitch-black and their position could not be seen until distress rockets were fired, when they could glimpse at that flash land but twenty yards distant, and only six from the vessel's floundering stern; but this solidity across a surge of sea in which few could hope to live.

Men – entirely on their own initiative – launched half-swamped boats, but to little avail. The *Iolaire* settled, and turned broadside onto the shore, breaking for some minutes the force of wind and water, and in this brief window one man made it ashore with a vital heaving-line made fast to the ship. A hawser was quickly stretched and by this rope around half the survivors escaped. Half of the rest, it may be reckoned, got ashore by other means; many would never be able to say how.

Even yet they were not safe. The coastline here is dangerous to the careless even in broad daylight, with crags and gullies in the eroding cliff. Inland, Holm is broken and boggy, rising and falling, thick in sections with impenetrable whin, making for fraught passage in the dark. And all these survivors were soaking wet – and most were barefoot – on a wild, wintry night with remorseless wind-chill.

The one matter that can never be agreed beyond doubt – apart, of course, from how the *Iolaire* ran aground in the first place – is how long she lasted before foundering. Survivor testimony (understandable, amidst terror and confusion and the manner time can stretch in life-and-death drama) varies wildly from about an hour and a half to as little as forty-five minutes. One witness from shore, who could see her mastlights and who was better placed for calm if solemn record, confirms what is probable; that she went down in barely an hour.

And, amidst fantastic, boozy confusion, rescue efforts from land had only just begun, compounding Lewis catastrophe with Stornoway shame.

Murdo MacDonald, Claoid Iain Uilleim, of 1 North Tolsta, was in one of the lounges and – like most – fell over when the *Iolaire* slammed into the rocks and listed heavily to starboard. There was only one door from this saloon, blocked by wholesale scramble to get out. Alarmed, Murdo smashed a window and crawled through: he had been going simply to jump into the sea, but took one look into the wind-lashed void and thought better of it. Still clinging to the gunwhale, he clambered back aboard and scurried to the boat deck.

The *Iolaire* had four boats – two substantial 'whalers', two lighter dinghys. Murdo MacDonald clambered with others into the starboard whaler. There was no officer to be seen and, men falling instinctively into the unspoken teamwork of Naval service, the boat was lowered. One young man, John MacLennan of Kneep, expostulated – 'It's only suicide, to lower the boats on a night like this' – but he was ignored. The craft had scarcely reached the boiling water when more men began jumping into it from the upper deck. 'I told the men to get out the oars and keep the boat off the ship's side, which they did,' Murdo MacDonald would remember. But someone on the boat deck – he could not see whom – was shouting that the boat be kept alongside. Not knowing what to do, its little crew obeyed; within a minute or two, under the hammering of the billows, it fell apart against the hull of the dying *Iolaire*.

MacDonald somehow evaded death, climbing up the after-grip and once, again, back on board. He ran to the galley, out to the port side – there was a boat on the davits, but he saw it was no use trying to launch it.

I then came down off the boat deck and went aft to the gun-platform. Then a heavy sea came and lifted the stern of the yacht on to the rocks, where it remained fixed. I saw some men on the shore who shouted to the men on the stern to throw a heaving-line. They tried to do so, but owing to the heavy sea washing over the rocks, the men on the shore were unable to reach the heaving-line when thrown to them. Then two other men and myself dropped onto the rocks about twelve feet below the gunwhale when the wave receded. One of them had a heaving-line with him. The three of us then held the heaving line and four or five men came ashore by it. After that, a hawser was hauled ashore and about thirty-five men came ashore by this. The hawser was kept in hand on shore by the men as they landed and was fastened to the ship when she sank . . .

But this was only his honest, fragmented bit of truth.

The *Iolaire* wallowed, swung about, leaned ominously further starboard, and for a precious moment her stern actually scraped the rocks of shore. Two men forward are said to have got clear by shimmying up her foremast – practically touching the low cliff – though it is recorded in no contemporary statement and may well be another dubious fragment of *Iolaire* folklore. A few abaft did jump straight onto these rocks and to safety; others leapt wildly into the sea, even as the ship was forced again from dry land to beat herself to death on the Beasts, her stern this time grounding.

John MacKenzie, 5 Port Voller, would live. As he prepared to leap, he saw Am Boicean, his cousin Alasdair MacKenzie, standing rigid by the rail. He shrieked to him; told him to get his hands out of his pockets and they would jump together. 'It's useless,' shouted the Boicean, 'I cannot swim . . .' John sprang.

And two ships sailed helplessly by, unseen. There was the little fishing-boat, the *Spider*, heading home after drift-netting off the Shiants. Her men had already seen the bigger ship, chugging past them off the mouth of Loch Grimshader. He thought it was the *Sheila*, James MacDonald would later attest:

> I noticed that the vessel did not alter her course when passing the light, but kept straight on in the direction of the Beasts at Holm. I remarked to one of the crew that the vessel would not clear the headland at Holm as it went too far off its course to make the harbour in safety. Immediately afterwards we heard loud shouting and then knew that the vessel was on the rocks. We were passing the Beacon Light at Arnish at that time, and could hear the shouting of the men as we were coming into the harbour. The night was very dark and a strong breeze from the south was raging and a heavy sea running. We were unable to give any assistance as we could not rely on our engine to operate in such a rough sea . . .

The *Sheila*, too, hastening from behind, could do nothing. She had left Kyle at least half an hour after the *Iolaire*; and as they neared port – Captain Cameron would later attest – saw the lights of a stricken vessel on the Beasts, could hear the uproar on board. But in such darkness he could see little, do nothing, could certainly not jeopardise the safety of his own command. Indeed, he feared his horrified passengers might sink the *Sheila* for him. 'Everybody came over to the starboard side – three or four hundred,' her passenger John MacLeod would say, 'and the Captain shouted. "For God's sake," he says, "Go to amidships! Don't turn the ship over!"'

'She listed pretty badly,' Lance-Corporal MacDonald would admit in 1968, 'but there was nobody – as far as I know – on board of the ship realised that that was the *Iolaire* on the rocks. We passed the Beacon on our port side, to get into the harbour . . .'

As the *Sheila* was made fast to No. 1 Pier, anxious voices rained from her decks. ' "Bheil thu sin, Aonghais?" – "Are you there, Angus?"; "A bheil thu sin, a Mhurchaidh?" – 'Are you there, Murdo?" ' John MacLeod, too, waiting to get off the *Sheila*, would recall besides the cries from the jetty. ' "Is my brother on board there?" "Is my father on board that boat?" I said, "No," I said, "No, no, none of your friends is on board this ship as far as we know." 'Cause we knew that they were on the *Iolaire*. And still, at the same time, we didn't know that anything happen . . .'

The anguished Captain Donald Cameron hurried away to report immediately that a ship was aground and that, in his firm opinion 'it was the Admiralty yacht with a large number of men on board.'

The *Iolaire* fought to summon help. On deck, a seaman from Shulishader in Point, Alexander MacIver, had found rockets; was now shooting them into the sky. But, fatefully, just another detail in the tapestry of damnation about the *Iolaire*, they were not explosive. Nobody heard them and many who saw those lights in town simply assumed it was a New Year thing, or perhaps Lord Leverhulme dropping in again; the new laird of Lewis liked to signal his approach from sea with some grandiloquent fireworks.

MacIver would not survive. On the bridge, after a seaman from Ranish – Angus Nicolson, who now had a Stornoway address – had screamed to the bridge to do it, Lieutenant Cotter now blew the steamwhistle; as boiler pressure sank and sank, he, too, began to fire rockets, rocket upon rocket, and light flares, red flares and blue flares. So rockets glowed high, did not bang; the flares blazed from fearful arms, but there was none in sight to see.

But what of the radio? Leonard Welch, twenty-three, from Malvern, Telegraphist of the *Iolaire*, gave most clear, sad evidence at subsequent interrogation. Off watch for most of the voyage, he was roused from his bunk at ten past one by Ernest Leggett, deckhand and quartermaster. Welch was on duty by 01:30. No signals came in – when the ship grounded, about 01:55, 'I tried to get into communication with Stornoway . . .' But nothing seemed to work. Alarmed, Welch examined all his equipment, and then raced to the engine room to see if all was in order. The dynamo was 'all right'. He returned to the wireless cabin and resumed his endeavours. The cumbersome radio equipment had probably been out of order for the entire passage.

Commander Mason appeared and ordered the telegraphist – rather unnecessarily – to send a distress signal. Welch declared his equipment was 'smashed' and then tried to rig an emergency set. He kept trying for half an hour, went to get a lamp, hauled back James MacLean – another of the crew – and made him hold the lantern as he fumbled on.

Then a huge wave smashed in on them, and Welch was quite disoriented, staggering down the ship as she started breaking up. He climbed to the bridge just in time to see his radio-shack – the 'deck installation' – washed away and reported to a faltering Cotter. 'Carry on,' said the Lieutenant vaguely, which Welch took as leave to abandon ship. (Many years later, Welch would greatly embroider this and other details.)

Another wave nearly ended the Telegraphist's hopes – his head hit something, and as he came to, clinging to the after gun-platform, he saw the decks now breaking up. He saw, too, a rope stretched out to shore, and men standing by it on deck, not willing to go, apparently 'stupefied'. Welch seized his last chance. He pulled off his boots, dived overboard, grabbed the hawser and toiled hand-over-hand to the shore through battering waves and sucking backwash. He had a 'hard struggle' on the beach, and felt so exhausted, and wanted to lie down. But to lie down was to die, and so Welch staggered on, following others. He came to a fallen man, well on in the Holm sward, and dragged him to his feet, and helped him along, and so they made it through ripping shrubs to the safety of a farm.

The conduct of the officers of this hopelessly overloaded, hopelessly under-crewed ship – as related in the scant sightings recorded in her last disintegrating hour – is unfathomable. Commander Mason was scarcely seen after the *Iolaire* struck; nor can his appearance have done anything to inspire confidence – emerging from his door moments after impact to call, 'What is that?' Welch – summoned twice to give evidence to the Admiralty Court of Inquiry – left the order for a distress signal on record, and that is the last we hear of Commander Mason, responsible master on a foundering ship with nearly 300 souls aboard.

Not many saw Cotter either – he seems to have remained on the bridge throughout – but he evidently fought to the last to win assistance from sea or shore. John MacKay, passenger, Leading Seaman from Shulishader in Point, had instinctively thought on impact, 'We've hit a mine'; and he, too, escaped his beleaguered saloon by breaking a window.

'It was a very dark night, but clear, with a very heavy sea on. I heard no orders given from the bridge. I was on the bridge when the rockets were sent up. I saw one of the officers there, but could not say if it was the Commander. I think the quartermaster was there too. The siren was blown three or four times. I and two others were searching for rockets, and the officer helped us and got the box of fuses for us . . .' The ship was being flung about so much, banging so hard on the reef, that men had constantly to hold onto support, leaning by a bulkhead, hooking an arm about a stanchion. Welch's main memory, besides, of the Navigating Officer, was more rockets, flares; Cotter firing a Very pistol, again and again, into the sightless sky.

Donald Morrison, Am Patch, moved for the lifeboats, assuming – like everyone else – there would be general muster-stations, disciplined crewmen, officers to supervise. And like everyone else, he witnessed only chaos. He saw two boats capsize in the sea before his eyes and had the presence of mind quickly to secure a lifeline. 'I threw down a rope to the boat that was under me. I and another lad then tied it to the *Iolaire*'s stanchions to secure it. Murchadh Iain Bhig, from Skigersta, later told me that it was on that rope he managed to clamber up again. A number of other men had also tried, but there were too many and they were too heavy for us to haul aboard – a terrible experience. The night was pitch-black and all you could see was the whirling white foam,' but – as the Patch added bleakly, sixty years later, 'you could *hear* plenty.'

Those aboard *Iolaire* wore only Great War naval uniform. Woollen, heavy, it made it all the harder to stay afloat; ashore, soaked, it is doubtful it afforded much insulation in wind-lashed conditions. Most dangerous were the sea-boots; practically all who survived had the wit to remove them (though most cut their feet badly ashore). In fresh water, one drowns quickly; it is absorbed through the lungs and fast, diluting and lethal, into the blood. But drowning in salt water is peculiarly horrific, advises local doctor and pathologist Robert Dickie, 'much less rapid, and the asphyxial element is greater with consequently longer survival time.' Indeed, many who died with the *Iolaire* would not, in the true sense of water inhalation, have 'drowned'. Dozens probably died of mechanical injury, pounded against rocks, and a large number would have died almost instantly from simple cardiac arrest. The phenomenon of reflex cardiac inhibition 'is associated with cold water entering the nasal passages and the mouth, and is more likely if the individual is under the influence of alcohol . . . hypothermia certainly has to be considered, especially as the naval gear could hardly be compared to a survival suit as used in the North Sea . . . it is often impossible to say what combination of factors – reflex cardiac inhibition, hypothermia, biochemical and haemodynamic upset, asphyxia – has resulted in death, and so "immersion in water" is an all-encompassing term.'

The general mayhem is evident in the recollections of James MacLean, the Campbeltown fisherman and *Iolaire* deckhand, who had seconds before impact been ordering ratings to ready their baggage. MacLean had been at the wheel of the yacht from 00:00 to 01:00 and, still around and still on duty, was sent by Cotter to ready the ship's hands for lowering anchor in Stornoway harbour.

'I was just returning to the promenade deck when the vessel struck the rocks. She took a terrible list and in about an hour's time sank. I was saved by getting hold of a rope and managed to scramble onto the rocks. Leggett, who relieved me at the wheel, was perfectly sober, and so was the Navigating Officer.' MacLean was adamant that no one else had been on watch but Cotter, Leggett and himself; that the grounding occurred about 'five minutes to two', and 'I was not on the bridge when the accident happened.' Under questioning, he admitted he had seen land on the port side 'before I went down at 1.25', a full half-hour before the *Iolaire* hit the rocks.

MacLean heard no orders given by Commander Mason or anyone else after the *Iolaire* grounded. He saw men lowering boats, but no orders had been given to this effect. And what did he do then? 'I was moving about to see what was going to happen.' The ship was already tearing up. A boat was launched, but quickly so swamped the few occupants hastily clawed themselves back onto the ship. The little dinghy drifted out to the full length of its painter. He saw four rockets go up, and met the Boatswain, who told him Cotter had just given orders 'to get the searchlight going.' Then McLean heard Cotter say – in words of remarkable ambiguity – 'Get into that boat – she cannot sink.' With a 'young rating' – it was 24-year-old Norman MacIver of 21 Arnol – James MacLean grabbed the painter and they heaved, but the dinghy would not budge. The rocks were about 'seventy or eighty fathoms' away.

MacLean was desperate to live. He urged MacIver to try and make shore by this rope. 'Well, you go first,' said the passenger. 'So I took my chance and pulled myself ashore by the painter.' The boat, awash to the gunwhales, was stuck in the rocks when he reached it. He swam the last few yards to land – 'it was difficult'. Others could have done the same, yes, 'but I saw nobody else,' save the boy, who soon came hand-over-hand to join him. Pressed, all MacLean could say of orders given was the second-hand instruction to get the searchlight working, and Cotter calling that 'the boat could not sink'. He had not seen Commander Mason, but overheard him crying 'What is that?' to Lieutenant Cotter. Others could have got ashore the same way, he maintained, but they were all at the stern on the other rope. He had called for others to come . . .

Under aggressive questioning, MacLean admitted that he had not gone to his boat-station on impact (though he was coxswain); he had not reported to the officer that the boat could not be lowered; he had made no attempt to fall the men in or control them in any way; and he had made no endeavour at all to get the libertymen out of the ship. We should not, though, be quick to condemn him for failing, freezing and terrified, to assume responsibilities the officers had quite abandoned. And James MacLean bequeathed one stark souvenir: his cheap pocket-watch, seized up by sea-water amidst the disaster, which he kept as a rusted remembrance for the rest of his life.

Another Tolsta man, John MacInnes, Iain a' Bhroga, 2 Hill Street, had a still more surreal encounter with Cotter. 'I asked the Lieutenant whether the tide was ebbing or flowing, and he told me it was flowing . . . I asked him if he had a searchlight to work. If we had a searchlight we could see where we were.' But Cotter answered that the dynamo was broken. Uncertain, awaiting instructions, MacInnes lingered for at least a quarter of an hour in the wheelhouse, as Cotter kept on firing rockets. And things were equally incoherent in the engine room, as Ernest Adams – Fireman; a stoker – insisted at the Court of Inquiry. When the ship struck, he 'stopped down' in the engine room and waited for orders with the Chief Engineer, Lieutenant Charles Rankin. He insisted that, almost immediately, the bridge rang down 'STOP' on the telegraph. But the Chief was just ringing back when the *Iolaire* shifted so quickly to starboard and he fell hard against her side. Adams stopped the engines as the shaken Rankin recovered. The telegraph, according to the Fireman, did not ring again.

Minutes passed, and then Rankin thought to order Adams to check the stokehold, to 'see what it was like . . . she was making water. I reported that to the Chief Engineer, and he and myself then went on deck.' Adams could not say when the dynamo stopped running. He insisted they had left it running. On deck, he helped get one of the boats out – 'the port dinghy' – but a fall failed and the little craft plunged uselessly into the sea. He retrieved its floundering occupant and tried to get to another boat, bumping into Leading Deckhand Charles Dewsbury. 'This is terrible!' exclaimed Dewsbury. 'Yes,' said Ernest Adams. He finally escaped by the rope. Dewsbury would drown.

There is evident confusion over power; Adams insisted the electric lights burned for an hour after the *Iolaire* struck. Griffith Ramsey, the other Fireman – who also survived – spoke dramatically of the flooding. When she hit, 'I was flung over towards the ladder. I shut the dampers up and I saw the water coming in through the port side just before the boiler, under the bunker . . . it was rushing in very fast . . . shooting in, in a good big volume. By the way the water was coming in it must have been a large hole.' Ramsay had the presence of mind to check the boiler pressure before he left – 'about 65 to 67 lbs'; normal working pressure was 84 lbs. He, too, got off by the rope.

'The sea was too heavy for any assistance to be got seawards,' John MacKay said dully a few days later, 'but if the rocket apparatus was on the shore it would have helped very much. If the ship had stayed a little longer on the rock I believe that most of us would have been saved, but when she went broadside to the rock, she sank.'

Help from land or water came there none; the only rockets any could see were those bolts and flares from the bridge, as Cotter fired on and on, thinking of nothing but securing such aid.

The pitiable tunnel-vision on the bridge of the *Iolaire* was matched – and hopelessly confounded – by like inflexibility at the Royal Naval Reserve Battery in Stornoway, the biggest RNR base in Britain. And drunkenness in town (though tiptoed past, then and now, by local writers) had its own role in the fiasco unfolding.

The first, it seems, to be aware of the *Iolaire*'s approach was William Saunders, Signalman at the Battery, on watch from midnight till four a.m. He noticed 'two steamer's lights making for Stornoway. One . . . passed into the harbour, but one appeared to have gone further east than she should have done.' Saunders would state that she 'stopped beyond Holm Point. The one that stopped, after a short interval, fired a blue light which I reported as the signals of a vessel requiring a pilot. That was followed by another blue light and I reported that . . .'

Lieutenant Robert Ainsdale was Officer of the Watch at the Battery. He telephoned Saunder's news – a 'blue light' at about five to two – to Rear-Admiral Boyle at the Imperial Hotel. Much can hang on a word. What Saunders had seen, of course, was a rocket – the word 'light' had that meaning, besides, in 1919 – and it was no summons for a pilot, but a frantic distress signal.

Precious minutes later, at 02:20 hours, Saunders saw a 'red light' and reported it as a vessel in distress. Boyle days later minuted, 'At 2.20 it was reported that vessel was sending up rockets – at this time the wind had freshened to a southerly gale, the atmosphere remaining clear. A drifter was sent out with Lieutenant Wenlock in charge. They failed to locate the wreck.'

Had they? Wenlock's own report differs. At 00:30, making ready to meet the *Iolaire* and serve as flit-boat for her passengers, he had headed out with his little command, HMD *Budding Rose*.

'At about 1.55 when on board *Budding Rose* I saw a rocket and proceeded immediately to investigate. I made for what I considered to be the position from which the rockets were fired and found a ship in distress on the Biastan Holm rocks, but was unable to render any assistance owing to the heavy seas running. I approached to the edge of the breakers, but found it was impossible to communicate with the ship in any way.'

He had sailed fearfully back to harbour and reported to Rear-Admiral Boyle at his hotel; on his orders, Wenlock then returned to sea and stood by until daylight. 'The conclusion I beg to bring to your notice,' wrote the anguished Wenlock, 'that I was on the scene of the disaster within half an hour of the first rocket being fired and that when the last rocket was fired I was as near to the actual position of the ship as was possible taking into consideration the safety of the ship and the heavy seas running at the time.'

One cannot fault Wenlock. But an eyebrow must be raised at the choices Rear-Admiral Boyle made as soon as he knew there was trouble. The beleaguered *Iolaire* was within sight of the Battery, and indeed the Imperial Hotel, her toplights clearly visible over Holm. Even on foot, a good many RNR men – taking the quickest route over the Sandwick shore – could have been on the scene in barely half an hour and could, even with nothing more than ropes and torches, have saved some lives. Boyle sent no one to the scene, did nothing to ascertain what was really going on. He fell back, like the male mind before and since, on the reassurance of available, costly technology.

The Rear-Admiral gave orders to rouse the local lifeboat, and to rouse besides HM Coastguard, who had a prized 'Life Saving Apparatus' – rocket-propelled breeches-buoy equipment, which weighed a whole ton and could only be hauled by road and, ideally, on a cart towed by a team of horses. The RNR Battery had no horses and it had not a single motor vehicle. And it was a quarter to three – nearly an hour after the grounding – before Boyle's order even reached Sub-Lieutenant C.W. Murray. There is high confusion as to when the Rear-Admiral sent word to the Coastguard.

Murray could do nothing but scurry on foot into town, and 'roused Mr MacKenzie, Royal Hotel, and told him to report to you,' he later recorded for Boyle. MacKenzie was Secretary of the Lifeboat Association. 'This occupied about twenty minutes, as the inmates slept at the back of the house, and, owing to the heavy wind, were unable to hear the door bell. In the end it was necessary for me to force an entrance. I then proceeded to rouse the caretaker of the Life Boat who lives opposite the doors of the Life Boat Station . . .'

But the caretaker had been in bed 'for three weeks' – Murray avoids the deadly word, 'unwell' – and simply handed the Sub-Lieutenant the key and gave him 'general instructions' to the home of the Lifeboat

Coxswain. Murray eventually found it and returned to light the lamps at the life boat station. Twenty minutes ticked by; then Mr MacKenzie, the coswain and three soldiers turned up. They told Murray flatly there were no other men available to crew the lifeboat. Murray could do nothing but return to the Imperial Hotel to report. It was now 04:30. It took him fifteen more minutes to wake Surgeon Owen. All the Sub-Lieutenant sought was a car, from one of the few hirers in town, at least to bring first-aid to the Holm scene. First he tried MacDonald, by No. 1 pier. 'I knocked at the door and rang the bell, but could get no answer. Then seeing a light in the middle window, top row, over the entrance porch, I again knocked and rang – still no answer. I then asked a passer-by if there was anyone in the house. He replied "Yes" so I returned, rang, kicked the door, threw stones at the upper windows, knocked at the lower windows, including those at the side of the house. Having no success I tried to force the door but it was too strong. I should say that I spent ten minutes trying to rouse the inmates.'

Not wishing to waste valuable time, writes Murray, he ran to Henderson's in Stag Road. Henderson had set up business hiring horses – then cars – and the family business survived into the last decade as a sad little liquor-store, a moral ending if ever there was one. But Murray – who did not know Stornoway well – found only the deserted garage. He struggled to locate the Henderson home, between darkness and widespread intoxication. 'Having knocked up ten or a dozen people, all of whom seemed utterly incapable of giving any coherent reply, I found the house and eventually found Mr Henderson. He opened a small window above the door and asked what I wanted. I replied that I required a car at once. He asked where I wanted to go and I told him Sandwick. He did not seem very agreeable, so I told him there was a ship ashore at Holm Point with 250 men aboard and I must get out to take assistance to any survivors.'

Henderson was unmoved, replying 'that he did not think that I could have a car, as his drivers were at home and probably would not come out on such a night – when I pointed out to him that men's lives were in great danger and delay might be fatal, he replied that he "didn't know" but "would see". I asked him if he would give me the key of the garage so that I could drive the car myself – this he refused, also my offer to buy one of the cars.'

Murray made again for MacDonald's, momentarily retracing his steps to tell Henderson to go straight to the Imperial Hotel if he got a car. Henderson promised to try. No car ever came to the Imperial, and none of the Henderson drivers were at work even by eleven a.m. on New Year's Day.

'As much time had now been wasted' – though Murray even tried to break open the MacDonald garage gates – he returned to the staff office to report again. It was now six in the morning. His next sally was to try, at least, to get a motor-cycle. He was still trying to wake MacDonald when he saw (joy of joys) a car roll up at the Post Office. Better still, the Postmaster was sober, coherent and concerned. He immediately put the car at Murray's disposal – it was the Ness mail-car, driven by Donald Murray, a stately rattle-trap with solid rubber tyres. He would drive all morning and then return to the Butt with his post-bags and, of course, with the bleakest news.

By 06:30, they were at the RNR sick-bay, to collect Surgeon Owen and Sick Bay Attendant Jones; also stretchers and First Aid Gear. After two short break-downs, they arrived at Sandwick about 07:15 – 'this being the nearest point to the wreck that we could approach by road . . .' The *Iolaire* had sunk four hours earlier.

The tale of Boyle's precious 'Life-Saving Apparatus' is still more pathetic. It sat in weighty splendour at the Battery depot itself, just half a mile from Holm Point as the crow flew. But – by his own testimony – it was 03:25 before Divisional Chief Officer F. Boxall of HM Coastguard was hauled from his cosy bed by an urgent message from Rear-Admiral Boyle.

'I told Chief Officer Barnes to call crew,' Boxall wrote on 6 January, 'and send for horses if not already done. I proceeded to the Battery arriving there about 3.40 a.m. Mr Barnes informed me that the horse had not arrived and that he could only get three of the Company. I then asked for sufficient men to proceed without horses, the Officer of the Watch giving nineteen men, leaving with LSA about 3.50 a.m.; the horse followed and picked up the party later.'

ACO William Barnes and party proceeded along the road, very slowly, hauling the Life-Saving Apparatus behind them. Boxall had sufficient wit to take one man aside and the two of them hastened instead by the shore, but 'owing to the darkness, boisterous weather and rough ground made slow progress and eventually by assistance of the man with a light managed to reach Stoneyfield Farm where there were several survivors, one of whom told me he was the last to leave the ship, and that she was totally lost.'

Barnes's account – written immediately on 1 January – varies only slightly. 'A message was received at 3 a.m. today from RA [Rear-Admiral] asking for LSA to be got out and taken to Holme Point [sic], 'Ship in Distress'. LSA was got out as quickly as possible; a number of men were requested from The Barracks as owing to the very rough nature of the ground the LSA cart could not be taken within a considerable distance of the place where ship was in distress and all the gear would have to be carried.'

On reaching as near the locale as they could with the cart, Barnes proceeded with two men ('one a native', he says grandly, 'who knew the place fairly well') and tried to find the *Iolaire*. 'I found a lot of wreckage at one point, but could not see any sign of the ship. I also searched along the coast as much as possible for signs of any survivors or bodies but found none. It was very difficult work searching the beach owing to weather; the seas were lashing the cliffs at some places so that it was impossible to get on the beach. After searching for a considerable time and finding nothing I returned to station with LSA thinking the men would probably be required at the Naval Base.'

But when, precisely, was word got to William Barnes? He was as emphatic at the later Public Inquiry as he was in this New Year's Day report that he was not called till 03:00. Lieutenant Ainsdale insisted he had received Rear-Admiral Boyle's order at 02:10 hours and had sent to Barnes and the Coast Guard immediately. The courier had returned at 02:30 and advised Barnes was coming at once. Pressed hard on this delay, Ainsdale asserted, 'The only delay would be when the Chief Coast Guard Officer sent for his men and could not get them. I don't know how long they waited for the horse at the crossroads; but so far as my department is concerned there was no delay at all.' As Dòmhnallach observed in

1978, there is a 'discrepancy of at least half an hour', but – if delay there were indeed – the culprit cannot now be identified; and the conflicting details in reports quickly penned later by exhausted men under Navy discipline should not greatly distract us.

Grotesque as all this blundering was, the people of Stornoway cannot be absolved of their own share of responsibility; everything in C.W. Murray's evidence attests to widespread drunken indifference. In any event, the matter was academic. No lifeboat could have reached or helped the *Iolaire*. And, to be of the least use, the ponderous LSA would have had to be on the shore and by the doomed ship within half an hour of stranding – which was impossible. It is certainly hard to believe very many men could have been saved by breeches-buoy equipment in the time available. Yet it is disconcerting to learn just how ill-equipped the Naval Depot in Stornoway was, even by the technology of 1919 – their transport woes are manifest, and they may not even have had electric torches: none are mentioned in the drama, and the LSA lighting defies belief.

John MacSween, a local man and the Coastguard detail's obliging 'native', could barely hide his contempt in later evidence to the Public Inquiry, as recorded in the surviving notes of a worried Admiralty onlooker. MacSween swore that he was

> . . . called at 3 a.m. by the W/op from Battery then at 3.10 by W/O Barnes. When he got to the Battery he saw the cart with the LSA apparatus [sic] at the gate, ready to go. MacSween then went to fetch the two carters and their horses. At about 3.30 he called the first carter on South Beach known as John 'Blue' MacDonald. Five minutes later he called Alex Neal the other carter. The horses were actually kept in a stable 400 yards from the Battery. When MacSween got back to the Battery the naval men there had hauled the cart to Sandwick Crossroads. At 4 am approx., the LSA cart started off from the crossroads and was taken another 1½ miles to within 200 yards of the wreck. It must have been nearly 5 a.m. By this time on MacSween's reckoning – (Far too late.)

> MacSween led Barnes to the wreck using one oil lamp with LSA, a carriage-lantern with a candle inside which formed part of the LSA apparatus [sic] which must have been near useless in the conditions. No better light was available nor rockets that would give light. They

saw wreckage but no men at all. They heard the shouts of the man on the mast but could do nothing. They returned to Holm Farm at about 5.30 a.m. On the beach they met two men from Sandwick. It was a very stormy night – a hurricane of wind and rain. Examined by Anderson MacSween agreed that if they had not waited for the horse they could have been at the wreck sooner. But MacSween says he had to obey orders (to get the horse.) Yet it was the sailors who dragged the apparatus the whole way on the outward journey. They would have been at the wreck by 4 a.m. The Divisional Officer of Coastguard is in charge of the rocket apparatus and the weight of the cart and equipment is fully one ton. Twelve men are supposed to be able to take the cart anywhere on an ordinary road.

Hugh Munro says they spent 10 mins. getting drag ropes on the cart before leaving the Battery (at about 3.40). They spent approx. 15–20 mins. at Sandwick Cross Roads waiting for the horse before carrying on without the horse. They went at walking pace dragging the heavy gear in the dark and wet.

MacSween was in front with his lantern as he knew the way to Holm Point. (What possible use could a horse-drawn LSA apparatus be in the case of a wreck 10 or 20 miles away? It was sheer awful coincidence that the *Iolaire* was wrecked so near the town that questions about the speed of reaction of the LSA could fairly be asked.)

That decision to linger at Stoneyfield Farm – accounts repeatedly confuse it with Holm Farm (adjacent but different) until daybreak may well have cost lives. There were still a few chilled, near-exhausted casualties floundering on the shore or lost in the Holm fields, and DCO Boxall can have been less than comfortable in completing his report. 'I waited some time intending as soon as the man could find his way to send for medical assistance and myself search round the scene of the wreck, but before any sign of daylight a Medical Officer with Sick Berth Attendant and another Officer arrived at the farm from the Base' – this was, of course, the conscientious C.W. Murray – 'attending to the survivors and sending some on by car. As soon as we could see at all the two officers from the base and myself proceeded to search, towards the wreck. I came across the body of a man who had apparently tried to reach the farm; the Surgeon attended to him at once but the man was already dead . . .' He had probably succumbed to exposure as RNR officers sat only yards away by a blazing fire, sipping sweet tea and smoking cigarettes.

As early as 03:30, five half-dead ratings off the *Iolaire* were made to report to Rear-Admiral Boyle at his office. 'Little information could be obtained from them,' he wrote later with staggering insensitivity, 'they appeared to have been washed on shore or else dropped from the stern when the vessel's stern was swung over the rocky shore at Holm Point – in fact they did not know exactly how they got on shore.' Even so, it was 06:10 before, as the scale of the disaster became apparent, Boyle could bring himself to alert his superiors.

from R.A. Stornoway
to Admiralty London
En Clair
596. REGRET TO REPORT YACHT 056 IOLAIRE WITH 260 RATINGS FROM KYLE TO STORNOWAY ON LEAVE GROUNDED ON EASTERN SHORE OF ENTRANCE TO STORNOWAY.
SOUTHERLY GALE BLOWING – AUXILIARY PATROL VESSELS SENT TO ASSISTANCE – UNABLE TO APPROACH IOLAIRE.
SEVERAL RATINGS HAVE REPORTED THAT THEY SWAM ASHORE AT VERY SHORT DISTANCE.
ROCKET APPARATUS DESPATCHED.
NO FURTHER INFORMATION AVAILABLE AT PRESENT.

Two hundred and sixty Naval ratings from Lewis and Harris had boarded the *Iolaire* at Kyle of Lochalsh. They had every right to be at ease, to feel safe. They found themselves instead, by colossal incompetence, shipwrecked but half a mile from home; they quickly grasped that the officers had not a clue where they were, nor the least idea how to retrieve the situation and preserve their lives. The libertymen also soon realised that no help hastened from land, nor could any emerge from the sea. The evidence is irrefutable: again and again we read, from witness statements, 'No orders were given', 'I heard no orders', 'I did not see any of the officers . . .'

No doubt a few panicked; some seem to have lapsed into shock, wandering about in a daze, clamouring in Gaelic – 'excited', said a scornful crewman; 'very upset', recalled a Tolsta survivor – and some slid further into a paralysis that cost them any chance of escape. But the real wonder is, perhaps, amidst confusion and dreadful conditions of wind and sea, that so many as seventy-nine men did live – because Hebridean passengers with grim purpose took charge, decades of local maritime experience realised to good effect. But bravery is by no means fearless. Their vessel 'was over the rock', Donald Murray would remember, 'she was pounded there by every wave, in such a way that she was shuddering from stem to stern. And us aboard were shuddering along with her. It was indescribable . . .'

Almost all the life-saving efforts in that dark hour were taken by the RNR passengers and on their own instincts. Alexander MacDonald fired the first rockets. John MacInnes coaxed the dull, overwhelmed Cotter into thinking of searchlights. Half of the survivors lived by the sober courage of one man, John Finlay MacLeod, who went out with the lifeline.

Shortly after impact, the thirty-two year old from Port of Ness got out to the exposed port deck. MacLeod heard no orders given for the men to do anything. He ended up in the drink as the solitary seaman placed inside a hapless lifeboat, but 'one of the ship's company got my hand and I got on board.' He beheld more desperation with boats, two – full of men – swamped by the ship's side. Then MacLeod took hold of a line – a long heaving line – and strode aft. He explained his purpose, fastened the thin rope about his wrist and in his grip, and unlaced his boots and flung them far into the dark. No, he would say afterwards, never thinking the question funny; he never found them again.

I let myself go, hands over the counter, like this – first wave I saw, I plunged into the water, and made for the shore, as *I* thought, but – oh – the waves had got up terribly at that time. And it wasn't far to go, but I didn't know how far, but during this time I just put the tip of my fingers against the rock. That was all, and this surge took me back further than I was before, and I knew that I'd never get a grip down there, so that there was only one chance. I tried to keep afloat and – see – I was then right under the counter again. And I remember – well, I was thinking, about the time – you know, now, that the rock's not far off, and just keep yourself afloat, and watch out for the waves, and don't take the first one, or the second, but watch the third one, for it's always the highest.

And, ferociously, defying cold and rock and every peril, John Finlay MacLeod swam round till that wave thundered in as he knew it would, and he rode along the crest of it, and was flung violently ashore.

He was badly bruised, cut; but he ignored it. MacLeod had to, or lose his life, for he was still in harm's way, helpless on a sloping face of knobby stone with a sheer shelf into deep water. At least there were ample handholds. He grabbed some rock, and the first backwash took him off his feet, and 'pulled me down a bit, but my hands held, and it was past – then I climbed up another yard. And while I was up about another yard, the line tightened – that was its full length.' He lectured himself, and the rope – *since I brought you here*, he thought, *I'm not going to let you get away* . . . He hung on through the next backwash, and the line – which had evidently snagged in his trail – came free, though worryingly short it was at that. He got 'up about two other yards, and sat down there, and started hollering to the people aboard, to come one by one. The first fellow that came was Iain – John Murray – Iain Help . . .

'I didn't know who it was at all, knew nothing about it. He got off all right. But in a minute or so, at best, I slipped down – there was too much weight on the line, between me and the ship, and they hollered to me that the two was on the line, and they were too heavy going over. I had the line round there, and I didn't really feel it much at the time, amongst the rocks there and what do you call it, *càit robh* . . . During the – all that time, there was five or six, something like that, came ashore on this line.' Moments later, as more men hand-hauled themselves over at the same time, their weight dragged John Finlay MacLeod within three feet of the deadly ledge; it is most unlikely even he could have come back.

All that was available, in a foundering ship full of frantic men, was one fat hawser. John Montgomery from Ranish was on deck. 'A dozen men' had already got to shore by the heaving-line. The first warp picked up was too short, too thick to knot. 'Somebody else passed a 2-inch rope over and we told them to drag it ashore. We got it ashore, and that was the only line we had . . .' It was too thick, not quite long enough, but it was the best available, made fast to the thin heaving-line. As more men came over, most joined the little party on shore who toiled to keep the rope as taut and secure as possible by their combined strength. Robert MacKinnon, the Harrisman, tried to make it fast to the rock, though by most accounts with limited success.

MacKinnon, like James MacLean, was one of the little band who had somehow made safety before this vital rope. The Harrisman had simply smashed through an uncooperative door to get to deck. 'At first I could not see the shore, but ten minutes afterwards when the rockets went up I could make out the land plainly. I made it up to the bow. She gave a great crash on the rocks and I saw the water rush up on deck through her buoys. I went to the starboard side, stripped off my oilskin coat and jumped into the sea. After a great struggle I reached the shore, but was washed out again on the receding wave. I then managed to swim towards a little bay and was lucky to come on a shallow place and so got on to a ledge of rocks and up out of reach of the sea . . .'

Storm; sea; the freezing wind; sodden clothing – everyone was on the brink of collapse, most sore or bloodied, most with flayed hands – and yet, with astounding discipline, they held on. New arrivals largely took their turn to anchor the rope until their strength failed. John Finlay MacLeod afterwards thought they were in operation for perhaps half an hour – 'It was there all the time until I left, but it was shifted to amidships . . . I was too exhausted. I could not hold on – it was impossible. The cold was too much for me.'

It was no light matter to come by this swaying, saggy line to land. And many who tried would die. For one, the *Iolaire* was in constant motion, listing and bouncing, her stern twisting, and all the time (no one ever forgot it) the howl and destruction of rock on iron. It took all the might of the drenched, chilled but wonderfully determined men on the Holm shore to keep the warp as rigid as possible; and then there was the raging sea – vast waves crashing in, slamming at men, at boys going hand-over-hand, or crawling upside down, doing whatever was necessary to cross these precious yards. Time and again, in a gurgling cry and briefest flail of limbs, a man was ripped away within feet of the shore, never to be seen again.

And meanwhile the ship herself was torn to death. John MacLean, a young seaman from Carloway, unnerved by the ceaseless screeching, grinding noises, found himself remembering lines from Longfellow, bashed into him at school: 'But the cruel rocks, they gored her side / Like the horns of an angry bull . . .' 'And, my goodness,' he would say, decades later, 'she was gored in many places along her side that night.'

One man now safe on land – a strong swimmer – was distraught. Donald MacDonald, of 13 Swainbost in Ness, was twenty-seven. His brother Murdo, twenty-one, had also been on board. But where was he now? MacDonald stumbled frantically about the Holm shore, crying, 'Am faca sibh Murchadh? Am faca sibh Murchadh?' 'Have you seen Murdo?' Then, without hesitation, he plunged back into the sea, swam back towards the *Iolaire*. He may even have made it. Donald's body was never recovered; Murdo's remains, found the following day, lie – alone – in the ancient cemetery of Swainbost. And Donald MacLeod of Tolsta, thirty-one, likewise returned to rescue his twenty-four-year-old brother; and they, too, died. They lie together at Cladh Mhìcheil.

Speaking many years later, it was still all too vivid to another Tolsta man, Donald Murray. 'Now while she struck – a while after she struck – everyone was looking out for himself – and I went up to the boat deck alone. They were lowering a boat there. Now I never thought of studying that there was no possibility of a boat surviving. In a state of excitement I went into it.' But the fumbling, frightened lads in charge of lowering made a botch of the falls and the boat was smashed 'into splinters' by a huge breaker. 'Now when the boat went to pieces I took hold of the ropes – you see? – because we was well acquainted with going up the mast and all that at that age, young, and everything. I grabbed the boat-hauler, went over her and right up and – aboard again. About the rest – there wasn't much that got out of her, they were all drowned in a couple of minutes . . . It was just that my time had not come . . .' He reached the stern, and remembered telling himself that 'the *Iolaire* would go down beneath my feet before I would risk moving off her again.' Dòmhnall Brus made no effort to hide how terrified he had been. 'From the stern

I could see the rapid succession of waves moving the stern backwards and forwards – tearing her. She was now holed and the sea was coming in.' By the time he had reached the right deck, John Finlay MacLeod had made it to shore with the vital line. Murray made close and fearful scrutiny of the fraught evacuation. At some point, wisely seeking a 'better lea' – more shelter – the lifeline was hastily shifted from the stern to amidships; again, men did this of their own bidding, despite subsequent Admiralty claims.

Increasingly alarmed, Donald Morrison of Knockaird was at least now gaining his night-vision. He had witnessed John Finlay's frantic swim, hoped and prayed. Now, painfully, through waves and mounting wreckage, with a few companions, he tried to reach this lifeline, floundering along steep, awash, cracking deck. The Patch was almost there when two huge waves overwhelmed them, and he was briefly stunned.

One young lad from North Tolsta, Donald MacIver, Am Bèicear – he was just eighteen – had a lifebelt in his bag, and the presence of mind to don it when the *Iolaire* struck. His escape story is typical. 'Well, I was young – I didn't know whether it was a rock or a torpedo or what had happened. The starboard side went down and the room was about – at about a sixty-degrees angle. The windows was up there, this was blocked – you couldn't get through the door; this had jammed. There was a rail going up, to the windows and to the door – the men was getting up on the rail as far as the door, but they couldn't open the door. The rest was hanging on to them. And then, next thing the whole lot was dropping down, back where they left. I got up this time and I just made for the windows. And there was a rail there, and I got a hold of the rail, and I put my feet up – out – first, and I got through that gap.' He smashed the window with his boots. Of all the Tolsta men in that particular compartment, only this teenage boy lived.

John MacLean from Carloway thought he was the twelfth man out and over by John Finlay MacLeod's rope. 'I was pretty lucky. The wave only caught me when I had my feet properly on the shore, and the wave only came up to here – I was one of the lucky ones at that time. But everyone that went on the rope, didn't get – they didn't get ashore. Some of them were thrashed away from the rope, and the first – a young seaman, fellow before me – thrashed on the rock. It would take some nerve for you to try and get ashore after. A lot of men, before they reached the shore from the ship, they were near half-dead. Sore feet and all . . . and we was never in the place before and it was pitch-dark.'

The fat knot, tied of necessity between hawser and heaving-line, was an additional hazard: not a few lost their grip on reaching it. But Donald MacIver had to try. 'I got down, I got a hold of the rope, and away I went. A lot of them that went down – lost . . . in fact the rope was too thick all together. I had a good hand then but it was a bit too thick. Washed ashore – next thing I was ashore, up on the rocks, bang. I didn't get a hold, I was out again, in again, I had come in, I think, three times, in and out. On the third time I got a hold. I said to myself, now, "If I don't get up I'm going to drown where I am." But I couldn't get up. The sea was back and fore all over me, and every time I tried to get a better hold. And I thought to myself, "Well, I'm going to drown here anyway." Still I could do nothing for myself. Somebody got a hold of me . . .' and so he made it onto blessed land, and within minutes had himself helped haul a man from Coll, Alasdair MacLeod, to safety.

Another young, young rating made it to deck and instinctively climbed into the rigging.

The ship was then heavily listed to starboard. I should say she was lying at an angle of forty degrees. There was a gale of wind with a high sea and it was bitterly cold. I was getting benumbed and was afraid I would drop off the rigging into the sea. I came down to the deck, my mind made up to jump into the sea and try to make the land which I knew was near although I could not see it, or tell which was the best way to go. After struggling about in the sea for some time I became entangled in a rope which brought me to a small lifeboat that had been swamped. I got into the boat which was full of water.

It may well have been James MacLean's boat.

After an interval, the length of which I had no idea, the boat struck against something and I put out my hand and found it was a rock, which seemed to be a small islet entirely surrounded by the sea. After some time I saw a light which seemed some distance away but my difficulty was how to get off my islet and towards the light. On trying to get up I found I could not rise so I made up my mind to lie where I was and await the return of the tide which I expected would submerge the rock on which I was. Covering my head with my oilskin, I stretched myself on the rock, and went to sleep.

Donald Murray was still on board. 'Well, the ship was going down fast then. I went on the rope at amidship. It was on the rope I went, so when I was on – when I reached the rocks and the wave was coming back I said . . . to the rope, I said, "Get at her, hold it fast," till the next wave was coming in. And I escaped.' He had thought through every last detail of this dash to safety.

There were teens of men – yes, fourteen men on the rope all at the one time going towards the shore. Now, once you were on the rope, you were safe enough if the waves were going towards the shore, do you see? The tragedy was when the waves were coming out one after the other and that is when they were swept off the rope. The first men to go on hadn't studied this. I was watching for a long time, as I had vowed to myself not to leave the vessel. I began to think that if I did go on the rope and had a good grip while the waves were coming in I would be amongst the boulders when the waves ebbed again. The returning waves were sweeping the rope bare – you cannot comprehend the ferocity of that sea. I was still watching – the sea was coming in under the rail. Many had now lost heart and would not go on the rope as they watched what was happening. I thought I had to take my chance of life or death when the next wave would be going landwards. When I heard the waves hit the outerside of the vessel I took my chance and got onto the rope and moved as fast as I could with the incoming waves along the rope.

When the wave returned outwards I took a death-grip of the rope. I felt the sea receding and the rope curving with its force and I knew this was the point at which the men were being swept off the rope. The next I knew the waves had passed and I was among the boulders. Still clinging to the rope, I moved as fast as I could through the boulders before the next sea would catch me. I was clear before the next wave came. It did not catch me. That's how I got away. My time had not yet come . . . *Dòmhnall Ghabhsainn* [Donald MacLeod, 58 Tolsta] came ashore. They say he went looking for his brother . . . and when he did not find him ashore he went out into the sea again to look for him. That is what was said . . . I cannot believe that anyone who got ashore would return. No, not even for your wife. It would be futile . . .

The *Iolaire* was in her death-throes now. 'I was one of the last to leave the deck,' Fireman Adams would later insist to *The Scotsman*, 'and almost as soon as I reached the land the yacht toppled over to port and disappeared – all but her masts. As she went down flames shot up. I am not sure where the flames came from; probably from the funnel.' Murdo MacDonald, Tolsta, also thought he had seen this. 'There was the report of an explosion and the funnel fell overboard. Then the yacht listed heavily to port and sank . . .' And Donald Murray painted it powerfully. 'I was only a while ashore when I saw a flash, coming out from the engine-room, a big flash – explosion – and, uh, she went down like a stone. There wasn't much aboard then but those that was aboard went down with her.'

Sore, groggy, Donald Morrison, Am Patch, had just regained consciousness – underwater, sinking with the *Iolaire*. He took the one chance he had. 'A rope came into my hand and I – and I pulled on it and I surfaced near – just beside the rigging of the main-mast, the mizzen-mast, whatever you call it. I got onto the rigging and I started going up and I looked behind and there was another chap coming up after me. Then I looked for'ard and she had toppled over onto her port side and there was nothing but two masts on the top of the bridge. She had toppled on to deeper water, you know. And a big sea came running and I looked behind after it passed and the fellow who was coming up after me was gone too. So I went up farther . . .'

Ashore, the rescue-party John Finlay MacLeod and Robert MacKinnon had organised so ferociously had the hawser – the lifeline – torn inexorably from their grip. There were still men crossing that line and they could hear their cries of despair as the *Iolaire* snatched them away; and then all was the roar of the sea, the collapse – somewhere – of plates and bulkheads, the bellows of the appalled party on shore and, from the night and from the sea, the terrors of the lost, and it was so *dark*.

They died below, behind stuck doors or maimed by injury, trapped and helpless in the ship that would be their tomb. They died, some, almost in an instant, hearts arrested by the sheer shock of cold, cold immersing water. They died in the pitch-black night, swimming frantically, swimming in circles, swimming in some instances in the darkness – it is whispered – towards Lochs; out into the Minch; further and further away from land. They died, clinging to wreckage, or buoyed by such combinations of lifejackets and belts as they could seize, still afloat, breaths shallower by the minute, tormented, as the cold reached their testicles, and then their limbs, and then their hearts. They died drowned by a slipped lifejacket, taking them about the hips and holding their heads mercilessly down. They died thrashing the sea into foam with flailing arms. They died struggling and sinking, entangled in rope, feet jammed, wreckage bashing their heads. They died by assault, by wave on rock, battered so savagely that some bodies recovered even within that New Year's Day could never be identified. They died beneath the *Iolaire*, coming up against her plates, banging and fighting to get out to air from her great iron bottom; or were slain by her, as she crushed them on the rocks in her rolling.

They died, some, because they could not move and were too terrified to make any bid to escape, and so died, really, of indecision: not because they made the wrong choices, but because they made no choices. They died as grey and bearded men, men with children, men with grandchildren. They died as men who loved women, boys who had never known women, boys who loved boys, boys who loved life and were in the flower of it. They died thinking of silly little things, wondering, annoyed, stoical. They died praying, or cursing, or in blank peace, in horror and wrath, in acceptance of the dark, the cold and of sleep, spitting oil and breathing, at last and suicidally, the sea itself, for surely the bitterness of death was past.

Would it never be light?

Barely conscious, as far up it as he could reach, Donald Morrison, Am Patch, heard shouting now, the wails of the drowning and the dying. He waited for it to stop. It did stop. 'And, ah, well, I was looking around and I looked up at her foremast and I could see other two men in the foremast rigging looking just the same as – just the same position as I was myself – one up as far as he could get and another one further down, just at the sea. To make sure, I started to shout to them, to see if they were alive. They shouted back – I didn't understand what they said . . .'

We were there, big seas were coming in, and she was rolling, the masts were going down, she was still rolling as though she had taken the ground, the masts still going down, the seas were catching her masts on the turn, they were going like that you know, and she went down, and then she went up – nearly on an even keel again. And that was making it worse, you know, the masts would go down and the sea would catch me at that angle, but – all I could think of there was, *you haven't long to go,* just thinking of – seeing death in every wave. I had a stay in each hand and I was sitting in the mast and both feet swung around the mast, you know, and I was sitting in that position there and I could see a big sea coming – I would squeeze myself onto the mast, making myself a bit of a shelter, you know. The mast going down, you know, and I heard another crash, and the big aerial she was carrying came right down on top of me. I didn't know what had happened, but when I got clear of that I looked for'ard again and the other mast had broken, and the men – the two men on the other mast had gone too – and there I was, all by myself, till daylight.

'I thought that I was going to die,' he mused long after, 'and that no one would know I was up there on the remaining mast.' But he was only twenty, Donald Morrison from Knockaird, and had come through much in the last hour and a half. He gripped yet, and tighter. Hours passed. There were men out there now – men on land; he could sense them, hear voices, there was a flash of dimmest light. The Patch cried and cried, but answer came there none. It might, he thought, be really the most sensible thing to let go, to end the intolerable ache and pain, to give up this useless resistance.

Something was silvering behind him, something just beginning on the eastern horizon. A verse came into his mind from nowhere, words he had learned . . . where?

Thig treis is furtachd thuc' o Dhia,
le fuasgladh an deagh àm;
Saorar iad leis o dhaoinibh olc,
oir chuir iad muinghinn ann . . .

The Lord shall help, and them deliver:
he shall them free and save
From wicked men; because in him
their confidence they have.

Something strengthened inside Donald Morrison. He shifted just a little on his fragile seat; and he clung that wee bit harder to his mast, the very last passenger of HMY *Iolaire*, against the night and cold and anguish, against the storm and the sea.

The Iolaire

(I)

The war over.
        In windy, flapping blue
kitbagged, slant-capped, we waited for our ship
to sail us home, there on the Kyle's roped quay
friendly with baggage, warm with singing and
with Gaelic gentle. Each had his own wound
(inward or outward): Jim the drooping arm
slashed by a Turk; John, the surfaced bones
pastured on greyness. Some with amazed looks
tested their smiles.

Tunics slap at our cheeks as we confront
(eyes dwindling seaward, grey on narrowing grey)
the year's last day, expiring westward now
like spent shell smoke.
        Tomorrow, the New Year.

Tomorrow confrontation with our dream
nourished by gritted passion while the guns
brutally rocked, recoiled, bruising the hand.

The quay sails outward. Whose are the lips whistling
that Gaelic tune? The packed grey brightens;
the catching wind abates. See how the harvest moon,
the barley-ripener, reddens the thatched grass.
Heavy and huge it broods (like God himself)
on the calm stooks, the windows whitening now,
the dancers dancing, the sea horizon-mounting.

        O my love my love
you who fed me when about the house
winged I went singing and the walls unfolded
ready for rising. You shall never starve.

Bread I will grow you, fish I will catch each day.
Till the world's end I'll keep you fed and happy.

That roar! Quick! What is that roar?
The quay steadies. It's merely the hooter. This,
at last, our ship.

. . . At last.

The clock tolls six. We're speeding seaward now.
Good wind from the south-east and the wake receding,
greening to starboard.
      And the songs beginning
for it's New Year's Eve.
      And the wind rising.

(II)

Shark-headed God snaps from his gardened calm.
His teeth bite cleanly and the fragrance burns
heavy with fruit and salt.
      The void mouth yearns!

What floods regret us! Sleekening wave on wave
what seamless tautened surface seaward turns
to the flagged sunsets of the plummeting grave.

Thus dreamt I waking in a bitter sweat
on almost the New Year's turn as the ship struck
her sudden death as cleanly as a clock.

(III)

The chart-house heaved. Out through the windward door
I fought a passage into lighthouse beams
(the harbour a mile distant not much more)
and there before us in the zigzag gleams
of waving light, the cliffs and the snarling roar
of feeding waters. This was no human shore.
This was the snapping beast of my broken dreams.

Brightenings, darkenings, leapings, call on call,
morsels of lifeboats clutched in the mounting showers.
The death-struck faces. Cliffs abrupt and tall.

Sailors on masts as thick as bees in flowers.
Boilers exploding. Jumps shoreward to that wall.
The milky suckings as the bodies fall.
Half-sleeping faces rayed with the light's hours.

Burning men plunged from decks: hissing they drowned
or rather they were there and then were gone.
A terror in my being so profound
that I was visibly shaking as might one
hugged by his married death when not a sound
ripples the sky and, over the dead ground,
steal, coldly-hued, the inhuman waves of dawn.

A sailor reached a ledge. He threw a rope.
I plunged straight downward gathering in mid-air
that slender prayer fastened to my hope.
I slipped, I steadily climbed, and, almost there,
I slipped again, weeping, stretching to grope.
I slid my childish death on that green slope,
embraced the dead, cursed them in my despair,

kicked at the living and was gathered up
safe to that ledge. Water, fire and air
roared in my mouth as, from a solid cup,
we drink a liquid which we can't compare
with any liquid we were used to sup.
We hauled them from the water drained and bare.
They climbed with gritted souls that blinding stair.

Two hundred drowned. The lighthouse flashed its light
a bare mile from home and there we lay
fished from the avid waters in the night.
The flood shone fatly south at break of day
assuaged by the New Year dead. From that safe height
we gazed straight downward. Only God could say
what dry and prayerless prayers we turned to pray.

Iain Crichton Smith (1928–98)
from *The Saltire Review*, Winter 1957

Chapter Five

'Grief Unutterable'

'An old man sobbing into his handkerchief with a stalwart son in khaki
sitting on the cart beside him, the remains of another son in the coffin
behind – that was one of the sights seen today as one of the funeral parties
emerged from the barrack gate. Another, an elderly woman, well dressed,
comes staggering down the roadway and bursts into a paralysis of grief
as she tells the sympathisers at the gate that her boy is in the mortuary.
Strong men weeping and women wailing or wandering around with
blanched, tear-stained faces are to be seen in almost every street . . .'

Glasgow Herald, 4 January 1919

Pounding, stumbling, anguished shouts . . . Mr Anderson Young started
awake, in immediate alarm, then anger. It would be some drunk, of
course, some fool who had thought to come first-footing on Stoneyfield
Farm – and what a night to be out in. And – yes – they had woken his
wife, stirring and complaining beside him.

They fumbled for matches, for a candle, and the lady of the house
swept downstairs like a galleon in full sail, a lethal glance at the clock
on the wall affording her yet further ammunition. And there he was – a
sailor, soaking, wild-eyed, disorientated, scarcely able at first to turn a
sentence, but within half a minute they had grasped there was a ship
wrecked but a few hundred yards away; that desperate men were battling
for life in the roar and spray of the night.

Mrs Young rallied splendidly. She took entire charge. She put the kettle
on. She roused the maids. She told her husband to make at once for
town, to get help. Bottles clinked as she unearthed the spirits, the port-
wine. And, with great presence of mind, she got more lights – lanterns,
candles – and put them blazing in every window.

In little groups, in pairs, some half-carrying each other, some crying, many incoherent, they stumbled through her door. The womenfolk were now in that exalted energy of those who feel sought and needed. They operated as efficiently as any professional emergency station, feeding, refreshing, drying, bandaging. Each man was quickly assessed. The worst they stripped, rubbed, and put to bed. As the house filled, Mrs Young called for blankets – more blankets – and shakedowns were fashioned in this corner and that. And still they came. She sacrificed, less than heroically, her man's tobacco, and about them all, amidst hot tea and aromatic smoke, with capable hands and will, Mrs Young engineered home, perspective, and something approaching normality.

Donald MacIver, the Tolsta teenager, cold and numb, bloated with all the salt-water he had swallowed, never forgot the welcome of this Stoneyfield Farm.

When I got in among the thorns, I was in an awful mess, my hands and face – came out of there, and I just happened to see a light. And I made for that light. Got to the door. Didn't have the sense – I went to the door, the wind was on it, southerly wind. They started shouting to come to the back. Before I got round the people of the house came out and met me. When I got into the house the place was full, they were lying all over the place. And on the table there was a basin of eggs. And I never was an egg man – I could never take an egg when I was young, and I just got a hold of the eggs and started scoffing them down, way over there, and I started putting up. That did me the world of good. Put up everything – I flooded the house. Soon as that was all over, I felt brand-new.

Even this minor atrocity was taken in their stride by the ladies. But, as Stoneyfield Farm was remorselessly overwhelmed by distraught sailors, Mrs Young spoke with the toughest men – the ones least hypothermic, those still in presence of mind and evident vigour, already rallied by a hot drink or a mouthful of ardent spirit – and encouraged them to make for town, where they might rouse those they knew and get a change of clothing.

So Donald MacIver and others walked into the streets of Stornoway, on New Year's morning, bare-footed and in nothing but the rags of what had been Navy duds – and these still soaking and the wind yet freezing, frost now silvering the ground. 'And I was a while on the road, Newton Road,' MacIver would tell, 'when all of a sudden I heard this commotion coming – this was the life-saving apparatus. The officer in charge looked at me, and he says to the boys, "Put him on the stretcher." I says, "Well, what for?" "Oh, you're going to the Depot." I said, "I'm not going to no Depot." He says, "Where are you going?" I says, "I'm going to the town." '

And he did.

Others had bypassed the farm entirely, and not all were so mentally strong, as Flora Boyd – then twenty-one – related forty years later. 'A sailor came to the door and walked straight in, because everybody had their lights on and their doors open by this time, and the boy just sat down and he didn't speak. So he was made comfortable and asked no questions, and sent to bed.' It was the quiet charity of an age most different to our own, and Flora Boyd would have her own grief by morning.

A young Harrisman would not stop in town at all. As soon as Alick Campbell got somehow ashore and grabbed some passing refreshment at Stoneyfield, the boy from Plocrapool walked. He kept on walking. He strode with near-demonic energy through Stornoway and south and out by Marybank, his mind locked on just one thing: the sanity of the high heart of Harris, his township, his home. And Alexander Campbell, fair as a girl, fit as a whippet, walked and walked. Early, he took a wrong turn, and by thin afternoon found himself on the West Side at Garynahine. He simply turned and detoured, and would not stop. He was in Ballalan by nightfall, and there they took him in, and he slept and slept. Next day, early, he was on the rove once more, shortly climbing the Clisham, winding through his own land, rough road and bog and rougher path, but he did not stop. He lurched through his own door at Plocrapool on the evening of the second day, and there Alick Campbell fell on those he loved and there, for the first time, the lad from Harris wept.

John MacLennan, from Kneep, was not much older. He actually belonged to the crew of the real *Iolaire* – then on the Clyde – but had come on leave for New Year. He had managed to crawl to life by John Finlay's rope. 'When I got ashore I was shoeless, as I had been resting and had taken them off . . . Reaching the shore, I fell into a bog and lost my socks – then I headed for the nearest house, where a huddle of injured people had gathered. I was injured with cuts to my chest, but I never let on to anyone. It was a frosty night and I walked from Holm to Stornoway.'

He was stumbling towards the Post Office near dawn when he heard a woman crying – an Uig girl whom John knew. 'It was Maga – nighean Sheònaid Chaluim Tharmoid. Maga had met two Uig men who had been on the boat . . . They'd mentioned I was on the ship and as they hadn't seen me since coming ashore, they'd come to the conclusion that I too had been lost. I got some civvies – they were clothes belonging to Peter, a brother of Maga. Then I went down to the Admiralty building, where a row of us were sitting outside on the wall. Admiral Boyle came and asked me who I was – he didn't know me in the civvies. Then he clapped me on the shoulder and said, "Thank God you're safe . . . come back here in quarter of an hour and I'll have the car ready for you."'

Thus the horror seeped into Lewis, even as the Youngs, the Boyds and others lit their candles against the night; an awfulness captured so memorably by the exhausted William Grant, penning his spare but powerful *Stornoway Gazette* report on 10 January.

> No one who is now alive in Lewis can ever forget the 1st of January, 1919, and future generations will speak of it as the blackest day in the history of our island, for on it two hundred of our bravest and best perished on the very threshold of their homes, under the most tragic circumstances. The terrible disaster at Holm on New Year's morning has plunged every home and every heart in Lewis into grief unutterable. Language cannot express the anguish, the desolation, the despair, which this awful catastrophe has inflicted. One thinks of the wide circle of blood relations affected by the loss of even one of the gallant lads, and imagination sees these circles multiplied by the number of the dead, overlapping and overlapping each other till the whole island – every hearth and home in it – is shrouded in deepest gloom . . .

There are not, today, many on the Long Island alive who can still bear witness to that New Year; those with us now, nine decades on, were then small children. Yet, like so many others now gone, the gate of 1919 lodged – as Grant foresaw – with abiding power in their minds.

The excitement of the evening and the bustle of women, the peats got in and the fires banked up, the snowy tablecloths and good things got ready, things they would not normally have – bacon, sweet biscuits, wheaten bread, things tinned, things that were treats; and a throwing open of kists and a rifling through stuff, men's clothes brought out to air . . . and the waiting, and the waiting, and the protests and questions, and then overwhelming sleep, to open their eyes to doubt, worry, more delay – then weeping mothers, fraught speculation, dread news, solemn elders.

My grandmother, six years old in Habost, Ness, remembered the air of festivity, the proudly laid table. She nodded off in a corner, forgotten. She was awoken by wailing. A mile up the road, Peggy Murray – only three – was as excited as her elder brothers. Normally her mother, Dolina, was *gu math smabail oirnn*, 'pretty firm with us', but on this night they were allowed to wait up until their father, John, came home. She sensed her mother's own happy anticipation – fussing with dishes at the dresser, putting the finishing touches to a special meal. "Bha duine às dèidh duine againne a' dol a-mach a shealltainn gu ceann an starain ach am faiceadh sinn an robh solas càir a' tighinn a-nuas mullach Thàboist. Dh'fheumainn mo 'thurn' fhaighinn còmhla ri càch, agus tha cuimhne agam air a sin, nuair a bha an 'turn' agam a' tighinn, cho luath 's a bha mi a' ruith a-mach 's a' coimhead, 's a' coimhead, 's cha robh càil a' tighinn . . . ' 'We were running out one after another to see from the top of the path but we were wanting to see the light of the car coming down the Habost brae. I begged my turn to go with each of them, and I remember that, when my turn came, how fast I ran out and watching, and watching, and nothing was coming . . .'

In Portvoller, Point, children had spent days building an enormous bonfire atop the low hill of Foitealar and by New Year's Eve – as Calum Ferguson's mother, Màiread, would tell him – it was 'as big as a barn, fat with driftwood, rubbish, dried bracken . . .' A great beacon of celebration would hail 1919. Màiread, 17, had her own thoughts – their father had been lost at sea, on active service – but her little brothers begged to hoist triumphal flags, as the neighbours had. Their mother had to retire to her bedroom, her eyes brimming. The boys finally raised a bamboo fishing-pole, not with pennant of scarlet or yellow, but a black shawl of mourning. 'There was a lot of silence in our house that day. Silence, though in our hearts we were weeping . . .' Their father, for certain, was not coming home.

Màiread was sent down to stay with her Aunt Marion at Aird, who was expecting her man – 'The Boicean' – and had six wee ones to handle besides. She found the household astir, exhilarated, as her aunt fussed over a stew. With much merriment, the children, one after another, were bathed, and a show of force was made about bedtime. Neighbours popped back and fore with little gifts for the imminent war hero – fish, crowdie, coarse cake sweetened with treacle. A clothes-horse was toted to the fireside; long-johns, a vest and a nightshirt warmed away.

Later, boys came by, begging for a little broth. There would be no bonfires tonight after all; it had turned so wet. They would kindle them tomorrow, all the better with the sailors home: their fathers, uncles, big brothers. Then an elderly neighbour, Norman MacIver, appeared, his eyes twinkling: the boats had sailed and they could expect Alasdair, Am Boicean, around three o'clock. Wasn't the seat already booked on his own gig? Excitedly, the women chatted on. Marion counted her savings – £27 17s. They could now afford to do up the house properly – Alasdair would do all the work himself.

Màiread at last nodded off in the warmth. It was the chill that woke her – hours on – the fire out, the lamp low. And her aunt was rocking back and forth on the milking-stool, back and forth, keening, grieving. 'Am Boicean will never come home . . . I am a widow . . .' The elders were there by six; Màiread's mother and her uncle shortly afterwards, and distraught women, faces glazed with tears. The wee ones awoke; pitter-pattered down to inquire. Worship was held. The Psalm was read, but not sung, the wise men mindful of an utter emotional limit.

Not too far from Stornoway, little Donald Morrison longed for his father's return. And most late at night 'we went to the end of the road over there, just to see what we could see, and we didn't see anything, but we heard the whistles of the New Year at twelve o'clock at night, and we saw some rockets – and we never thought, you know – it was just, uh, taking in of the New Year. But evidently it was something else. It was the rockets, the distress-signals from the *Iolaire* . . .' Now it was day, 'and I can remember in the morning an old Christian woman coming in. My mother was up and she said something to my mother. She knew, from what she told her – she knew that something was wrong. Well, my mother fainted . . .'

Marion, my cousin Mòr, is today almost the last close witness to that *Iolaire* New Year. She was four years old, in Earshader, Uig, awaiting the return of her father, Kenneth Smith. 'I very clearly remember being told that Father was going to come home, and he wouldn't be going away again. And I can remember my mother airing his clothes. There were no buses then, and very bad roads, so the men would have to walk all the way home from Stornoway, and their clothes would be soaked. So she had clothes and underwear and so on airing in front of the fire, the clothes that he would wear in civilian life.'

Small as Mòr was, as time dragged she began to feel something dreadfully wrong. It puzzled her. But her strongest memory is most female, territorial. A woman she scarcely knew, a neighbour, bustled into the living-room, and gathered up all those clothes still ready by the fire, and vanished into the next room. *That's cheeky*, thought Mòr darkly. *This isn't your house* . . .

It was an uneasy morning in Lionel. Peggy's mother was preparing a very quiet breakfast, and that was when wee Norman from the police-house rushed in. It was just three houses down on the other side of the road, Peggy would say, and he was ages with her older brother, another Norman – seven years old, and they played together. Well, men had gathered by the police station, talking, and Tarmod Plò was just seven, after all, and he had an awful stutter – a stammering – and in he burst, running like life. 'Tha . . .tha . . . tha na fireannaich aig c-c-ceann na sràide ag ràdh gun tàinig fios gun deach an t-soitheach anns an robh Iain a' tighinn air na c-c-creagan bho raoir 's gun deach na d-d-d-daoine a bhàthadh . . .' 'The-the-the men at the top of the road are saying they got news that the boat John was coming on went on the r-r-r-ocks last night and they are all drowned.'

Mrs Murray was by the dresser. And she clutched it, and paled, and she said to the elders already in the door, 'Is it true what he said?' and they said, 'It is true . . .'

She had been Mrs John Murray: somebody. She was now the Widow Murray, alone with tiny children: nobody. And that night, or some night in those nights, she dreamed, and met her John in her dream, and cried, 'How am I going to manage?'

Well, that's what I thought, too, when I heard the bell . . .

By greasy New Year afternoon, every triumphant flag on Lewis was down. The bonfires in Portvoller, at Aird, and all over the island were never lit. They would lie there, rotting on the moor, for years to come.

There is an old Gaelic proverb that, between Christmas and New Year, the daylight lengthens 'as a cockerel's step'; but though 21 December – the midwinter solstice – is the shortest day, and the sun thereafter sets some minutes later each evening, at northern latitudes sunrise continues to be a little later each morning, too, not starting to retreat until some days into January. It was still dark as the first alarmed people hastened to the sward of Holm, the shores of Stornoway, on New Year morning, but already – all along the sweep of gravelled beach by Goat Island and up by Sandwick and to Holm itself, the evidence of catastrophe was piling. Wreckage, baggage, pieces of smashed lifeboat and empty lifebelts, assorted personal effects and – rolling in the wash or bobbing in the ominous, wrinkling shallows – bodies and bodies and bodies.

Details haunted Anna MacAulay in 1969 – dolls and tiny boats and little teddies. 'It was a beautiful morning after a terrible stormy and dark night, and lying on the shore were the bodies of the sailors that were washed ashore, and the toys which they had been longing to bring home to their children, and hats belonging to the sailor-boys, strewn along the beach. I can remember that as vividly as if it was today. It was a tragedy. The whole island felt it, the sadness – you could feel it in the air. The tragedy of the thing. Although they lost many men during the war – and we fought the war in the sea – the sea was their livelihood. The sea gave, and the sea took it away . . .'

Yet – even allowing for a village bard's dramatic licence – the recollections of Murdo MacFarlane, seventeen that New Year's Day, linger more than most. He was with a group of fisher-girls, trying to make sense of the fractured news, when they resolved to leave their work for a little and head round the bay by the Lower Sandwick road to have a look.

At the foot of the brae they stopped, and they said – in Gaelic – 'A Dhia na tròcair, ach dè tha seo??' 'God of mercy, what is this?' So they didn't go any further. I carried on. The first victim was a sailor with his back to this wall here – drowned, of course. You could say that the cruel sea and the cemetery were competing as to which of them would claim his body.

At the end of the wall here, on this green patch – there's a fence on it now – the captain was there, with two lifebelts on. Drowned, of course. The man after me said, when he saw the two lifebelts on, 'Cha do shàbhail sin thu.' 'That didn't save you.'

But, although the wind had calmed down, the sea was still turbulent. And sailors were coming ashore. The breakers were raking the shore and you could see drowned sailors on the crest of the waves. They were like war-horses racing – galloping – from the field of battle, with their riders dead in their saddles.

Now at the end of that village there, there is one house – her name was Maidheag Ailein. I think it was her only son. He had come through four years of war, with his own house on top of the rocks, and here he was . . . drowned at his own door . . .

It was the body of twenty-four-year-old deckhand John MacAskill: born at Lower Sandwick, washed up drowned by Lower Sandwick, after all his service in the Great War. Flora Boyd was his fiancée, proud of the beautiful diamond ring he had bought for the engagement.

But no one ever saw the body of Commander Richard Mason – or, for that matter, Lieutenant Leonard Cotter. They were among the many, many never found, already dissolving in the cold Hebridean sea. If MacFarlane did see a dead officer, it must have been one of the engineers.

So day dawned, this beautiful day, thin light crawling into it, and to three men, against the longest odds, it would bring the gift of life renewed.

First, the boy. On the rock he had decided so placidly would be his death-bed, a sailor-lad stirred, sore and cold. 'Day was breaking and I then found that I had been on the shore all the time and the . . . farmhouse was not far away.' Somehow he dragged himself there and was just being made most comfortable, quite the pet of the ladies, when Surgeon Owen appeared and ordered his removal to the Battery sick-bay, to the boy's ill-veiled chagrin. But, he would insist to William Grant, 'I am quite all right now.' We cannot now know his name, but amidst such death, and in shades of 'The Piper at the Gates of Dawn', he had been quietly spared.

There was, besides, Am Patch. In the high swell and what was still a determined wind, the wreck itself was now coming into focus, pathetically close to land, one mast rearing from the waves at a crazy angle – and there, horrified onlookers could clearly see a living sailor, so near they could shout to him, but quite beyond their reach and, unless this wind and swell abated, beyond any rescue.

Donald Morrison had been resigned to die, but his verse had cheered him, and as day brightened and the sea began at last to settle – unpleasant rather than murderous – he felt 'a lot calmer; I felt hopes rising again. Then I saw a trawler and small boats approaching, but the sea was still not calm enough for them to get near. I remember thinking that it was too dangerous and that they would also sink.'

The Beasts could wreck again, and the hulk of the *Iolaire* was a new hazard besides. But the men now converging on Donald Morrison's perch were determined he, at least, would live. 'There had been a report sent in that there was a man clinging to the mast,' Lieutenant Wenlock later told the Fatal Accident Inquiry. 'That was a little after nine o' clock. It was light that morning at nine o' clock or a quarter past. I immediately took the motor launch *Spense* and with a volunteer crew went out to rescue the man.'

'The sea was still pretty rough at daybreak,' the Patch would relate, 'but between then and ten a.m. it seemed to subside a bit. The first boat attempted to come near where I was hanging on, but it had to turn back. I thought then I'd be safer where I was than trying to get to that boat.'

By the time Wenlock reached the wreck in the *Spense*, the whaler *Norqual* was pouring oil on the water near the wreck. These days they would have Greenpeace on him, but sailors have known for millennia the difference a spread of oil can make on unruly seas. The slick suffocated the wicked little waves that menace small boats, and nervous optimism spread with it.

Then the other boat tried and came nearer to my left side. They shouted to me, asking if I could clamber down. They approached to within hailing distance, and the officer in charge, he shouted to me, 'Can you come down?' There was a stay from where I was, near toward me, going down into the port gunwhale, and the port gunwhale was under the water, and I shouted back to him, 'Sir, if you can take the boat in between the mast and the stay, this stay here, I'll come down.' So that's what I did. He came in – came along with the boat – and he turned her round, and backed her in between the stay and the mast, and when I saw her coming just underneath me I jumped onto the stay and I came down hand over hand and I landed right down on the stern of the boat. And the officer put his hand out and he pulled me in. In a brief moment of time, I was aboard and they were so happy to see me.

The Patch, though near to physical collapse, cheered a little. He was still a young man with a young man's black mischief. 'Can I take an oar?'

They were quickly alongside the *Spense*, where Donald Morrison was helped up by willing hands. 'On board, they removed my wet clothing and gave me dry garments and hot tea,' and so they turned for the haven at last of Stornoway harbour.

Contrary to quick Admiralty spin, Lieutenant Wenlock did not personally rescue Donald Morrison (though he did hold a medal with bar for life-saving, as Boyle made sure to tell the Press) and it is by no means clear that the men of the little skiff that finally retrieved the Patch, and at some risk, were Naval chaps either. Though he asserted on several occasions that he had retrieved Morrison 'with great difficulty', Wenlock himself studiously avoided going into any detail. No doubt the exhausted Lieutenant hovered close and supervised operations.

The Patch had the last quiet laugh, perhaps, in his final years, talking with a certain determined innocence of 'Lieutenant Willock'. For now, inside dry clothes and outside warm tea, Donald Morrison insisted on climbing the steps of the quayside himself. He soon recovered at the Lewis Hospital and, though hard news awaited him – his brother, Angus, was gone and his remains would not be found – two things no one would ever take from the Patch: that mysterious verse, which had succoured him before dawn, and a wry footnote in history. Of the 284 men who had been aboard the ill-fated *Iolaire*, as Donald Morrison would tell admirers to the last decade of the century, 'I was the only one to step ashore on Stornoway Pier.'

Many, many would never set foot on Lewis again, and as bodies and still more bodies lapped in the Stornoway tide, Rear-Admiral Boyle – now fully aware of the scale of the disaster and, besides, that all the *Iolaire* officers were lost – focused on his immediate problem: the recovery of the dead. He summoned Lieutenant Townend, who had not long been roused in the small hours at his Church Street billet by a furious landlord – 'Why are you sleeping there when Lewis seamen are drowning on the Beasts of Holm?' Townend, shocked and groggy, had dressed immediately and reported to base.

Stornoway had a town mortuary, on Cromwell Street Quay, but all realised it could not cope with disaster on this scale. 'In the morning, the base Commander ordered me to take charge of all the bodies brought in, pending identification and awaiting collection for burial,' Townend would tell Calum Smith in the 1950s. 'It was a situation on which there was nothing in the *Manual of Seamanship*, and with which I simply did not know how to cope.'

Lieutenant Townend approached the Master-at-Arms, explained the situation, told him to forget the youth was an officer and just to give him man-to-man advice. Disconcerted, the fellow snapped to attention, saluted, and caroled, 'Whatever you say, sir.' Townend pressed him again; back came the same void of initiative. The officer's patience thinned. A solution came to him. 'Clear out the ammunition store!'

'But, sir . . .'

'Clear out the ammunition store! That's an order!'

It was the largest single apartment in the RNR Depot at the Battery, hard by the shore, readily accessible to bereaved families. As perspiring men carried away cases of bullets and shells, Townend gave urgent thought to arrangements. He secured tarpaulins, sturdy brown paper bags and chalk, and awaited the carnival of the dead.

The first survivors, the toughest and the least physically damaged, staggered into their own townships within hours, only to find themselves beset without respite by desperate relatives demanding any word of a husband, a father, a brother, a son – often the moment when these men finally broke down, as John MacLennan remembered,

'The car took Tuireag [Malcolm MacRitchie, 7 Aird Uig] and me to Callanish. Duncan MacRae's motor-launch took us to Tràigh Sheanais in Reef. Tuireag and I walked back to the Kneep and we met my sister Hannah at the top of the hill. She was so happy to see me. I never said anything. I went into the house. My mother was in bed and I went to her bedside. She said, "Iain, what happened?" "Nothing happened," I said. "Something happened – I've known for a long time ago that something dreadful was going to happen." I made it home on New Year's Day. The telegram with the news of the loss of the *Iolaire* wasn't put up in the Post Office in Uig until the following day. Somehow word must have got back to Uig because Càm and An Gobha came to the house, and as I hadn't seen either of their sons since I came ashore, I feared the worst. I went to hide at the other end of the house and I wished I hadn't come home so soon. Going to face those men knowing their sons were drowned, but I couldn't tell them that . . .'

That same New Year evening, the Tolsta men took the last miles to home. They 'got down to Tolsta in Coinneach Ruadh's gig,' Donald Murray recalled. 'It was up in Stornoway meeting his son, Dòmhnall Red. Ciorstag had gone up to meet her brother [Donald MacIver, 38 North Tolsta], but Dòmhnall was lost and she was returning without him. She brought home the boys that had been saved – there were five of us. I still remember that it was my sister Christina who brought me in from the road. I had seen Dòmhnall Red after the *Iolaire* struck, but I do not know what happened to him. I don't think he went on the rope, but then perhaps he did . . .' So they reached their native village, as people emerged from fraught evening prayer meetings.

Like everyone else, Donald Murray's home was quickly swamped with irrational inquiries; in a day or two, like the others who had lived, he had to begin the round of grieving homes. He had over seventy years more to spend in this intense little community, living alongside many bereaved by the *Iolaire*. Of the sixteen North Tolsta men who had boarded at Kyle only these five made it home. The casualties included Kenneth Campbell, Peatair, of 54 North Tolsta. Of these seven brothers, this impish young man was the only one aboard HMY *Iolaire* on 1 January 1919. The other six all survived the First World War.

Shawbost, too, saw such scenes. 'A dejected-looking sailor, wandering in a forlorn fashion in through the village, stopped and asked me in a perfunctory fashion what my name was,' Hector MacIver would record decades hence, 'and immediately seemed to lose all interest in his question and its reply . . . I looked at this survivor and was so appalled by the look on his face that the wreck was banished from my mind. On a fence, fifty yards away, a crimson bedcover was torn and flapping in the wind. The thought came immediately into my mind, "Are there crimson shrouds for sailors?"'

His schoolfellow, Calum Smith, just seven years old, was likewise disturbed.

A group of us were standing in front of one of the croft houses in North Shawbost when a sailor in uniform came trudging wearily along the street with his head down, and as he went past one of the boys called, 'A Mhurchaidh, an tàinig m' athair-sa a-raoir?' 'Murdo, did my Dad come home last night?' We all thought it strange that Murdo didn't lift his downcast head, look round or make any response . . . When the news did come through, my recollection is of everything suddenly going very quiet, of women talking in hushed voices: it was as if there was a feeling that noise would be an offence to the dead.

One of my most vivid recollections is of sitting in a neighbouring house the following day as one of the survivors sat on a stool in front of a large peat fire, with his trouser legs rolled up, while his mother knelt at his feet, bathing them in warm water, drying them very gently with a heated towel, and then smearing Vaseline on the cuts and abrasions that he had sustained clambering up the rocks at Holm in his socks. It was a time of desolation, and of a grief that still, after more than eighty years, touches those who remember.

Lewis gathered her dead.

Late this same New Year's Day, the gaily painted blue-and-red carts rolled in from the country. Friends and relatives already combed the Stornoway shores, with overwhelmed RNR men besides; and little boats converged by Holm, toiling with ropes and grapples. Not a few men, believing their kindred safe, came over for the remains of a neighbour's boy – to recognise the drowned features of their own son, or uncle, or brother.

Thus it went for days on end, until only fathers and sons, brothers and friends, were left to comb, week upon week, the shores of Holm, Stornoway and Point and North Lochs, hoping for a loved one and braced, as time passed, for horror and for something scarcely now recognisable as a man; searching for a body they could bury, begging for a finality, a closure, that in many instances would never come. Decomposition in salt water is slow, but horrible. The exposed parts of the body are soon gnawed by sea-life and, if afloat, fed on by birds, rapidly disfigured. A ghastly 'washerwoman's hands' wrinkling strikes sodden skin; bloody froth bubbles from the airways of the drowned.

Still they toiled. One old man went down each day in his best Sabbath clothing, in respect for the heroes and of other men's sons, week upon week. The commitment many gave in time and energy – to say nothing of real steel as the days stretched and those they recovered looked the more horrific – commands admiration, especially when so many engaged in the labour had lost no one.

As bodies were gathered, they were placed gently in horse-drawn lorries and borne to the Battery and the cleared ammunition store to be laid on the floor. Most still wore identity-discs, but it would take time to collate numbers, telegraph them to authority, put a name to a face. For now, tarpaulins covered them, and Townend ordered that all personal effects be removed from the pockets, logged, wrapped in one of his brown paper bags, and that bag precisely numbered: the same code was then scrawled on the dead sailor's boots – or, when there were no boots, a ticket pinned to the sodden uniform. ('For months after it was all over,' Townend would say, 'I saw in my dreams rows of naval-issue boots with numbers chalked on their soles.')

And still too many did not know – could not be sure – who, precisely, had sailed on HMY *Iolaire* from Kyle of Lochalsh.

'A visit to the scene of the disaster on Thursday revealed a heart-rending sight,' William Grant would record for his newspaper. 'After leaving the road as it turns into Holm Farm a walk of a few hundred yards brought one to the green sward overlooking the wreck. Here were gathered no idle sight-seers, for all had come in quest of the remains of near relatives.

On the grass were laid out the bodies that had been recovered from the sea, and below the crews of eight row-boats proceeded in silence with their work of dragging round the wreck. At very short intervals the grappling irons brought another and another of the bodies to the surface, and the crews proceeded with them to the ledge where they were being landed. Here they were placed on stretchers and slowly and laboriously the labourers clambered up with them to be laid out reverently on the grasslands above. Scarce a word was spoken, and the eyes of strong men filled with tears as the wan faces were scrutinised with mingled hope and fear of identification.

The remains, as they were recovered, were brought to a temporary mortuary at the Naval Barracks, where relatives of the missing men from all parts of the island gathered. As the bodies were identified they were handed over to the friends, and the little procession of carts, in groups of two and three, each with its coffin, passed through the Barrack gates on their way to some mourning village for interment . . .

'For days after the disaster,' recalled James Shaw Grant six decades later, 'I was not allowed out alone, in case I wandered down to the temporary mortuary established at the Naval Barracks. My brother, being older, had more freedom. He saw for himself the horror of it all, and brought back from the beach a fragment of the *Iolaire*'s handrail, which I still have.'

But no one captured those hours more powerfully than Donald MacPhail, an ageing schoolmaster from Bragar, interviewed by Fred Macaulay in 1959 for a landmark documentary on Gaelic radio.

Cha robh annamsa ach an gille òg aig an àm – bha mi seachd bliadhna deug – anns an Àrd-sgoil an Steòrnabhagh, agus tha cuimhne mhath agam, Là na Bliadhn' Ùire. Fear a bha an ath-thaigh, an ath-dhoras

dhomh, thàinig e nall – tarsainn na mòintich, tha mi smaoineachadh;
cionnas a fhuair e air chan aithne dhomh – ach bha e mar duine às a rian.
Agus bha an fheadhainn a chaill an cuid anns a' bhaile – na màthraichean
's na mnathan – bha iad a' tighinn a-staigh a dh'fhaighneachd am fac'
e sealladh air Dòmhnall, no air Aonghas, no air Eòghainn; 's cha robh
e ach ghan coimhead agus na deòir ri tighinn a-nuas air na gruaidhean
aige 's bha dà fhacal aige, tha cuimhne agam air a sin, bha dà fhacal aige
a bha e cantainn tric – 'Good God . . .Good God . . .' mar gum biodh e
air grèim fhaighinn air na facail sin air bòrd; 's lean iad ris an inntinn
aige, 's cha robh facail aige ach sin fhèin . . .
   Ach 's e gnothaich duilich a bh' ann dhan fheadhainn a bha feitheamh –
mar a bha am bàrd a' cantainn, 'bha an dachaigh a' feitheamh riutha, blàth,
's gach nì mar a b'fheàrr air dòigh' – a h-uile dad ullamh airson biadh
agus aodach airson an fheadhainn a bha tighinn, carthannas agus blàths,
na dàimhean a bh' aig an dachaigh, agus naidheachd thùrsach a' tighinn,
nach ruigeadh iad tuilleadh.
   Dh'fhalbh mise a Steòrnabhagh; tha cuimhne agam, 's e deireadh
oidhche a bh' ann, le cairt 's each, mi fhìn 's gille òg eile – chan eil fhios
agam nach robh dithis eile còmhla rium agus athair fear dhe na gillean
a chaidh a chall; agus chaidh sinn sìos dhan Bhataraidh far an robh na
cuirp air an cur a-mach ach an aithnicheadh daoine iad, 's tha cuimhne
agam gu robh ticead orra – Liùrbost, agus Siabost, agus Tolastadh. Agus
am fear à Siabost a chaidh a-null còmhla rinn, bha mac leis ann, 's tha
cuimhne agam, 's cuimhne agam gu robh e cho brèagha 's gun canainn
nach robh e marbh idir, am fiamh a bh' air aodann – tha cuimhne agam
air sin cho brèagha fhathast.
   Chaidh athair air a ghlùinean ri thaobh agus thòisich e toirt litrichean
às a phòcaid, 's bha airgead, tha cuimhne agam, airgead-geal agus airgead-
pàipeir, tha cuimhne agam, airgead-geal agus e air fhaighinn na phòcaid
agus na deòir ri tuiteam air corp a mhic, 's bha mi smaoineachadh gur e
sealladh cho tiamhaidh 's cho duilich 's a chunnaic mi riamh; 's cha robh
sin ach aonan de mhòran a dh'aodadh sinn fhaicinn anns a' Bhataraidh
an là bh' ann an seo – 's lathaichean às a dhèidh.

I was only a young lad at the time – I was seventeen – in the high school
in Stornoway, and I remember it well, New Year's Day. A man in the
next house, next door to me, he came home – trekked across the moor, I

think; how he got home I do not know – but he was like a man out of his mind. And those people in the village who had lost men – the mothers and the wives – they were coming in to ask if he had seen any sight of Donald, or Angus, or Ewan, but he could only look at them and the tears coming down his cheeks; and he had two words, I remember that, he had two words that he said often – 'Good God . . . Good God . . .' as though he had caught on to those words on board and they had stuck in his mind, and he had no other words.

It was a very sorrowful business for those who were waiting. As the bard said – *'Home awaited them warm, and all was best prepared'* – all had been got ready; food and clothing for those who were expected; friendship and warmth, the families at home; then the awful news that they would never come.

I left for Stornoway – I remember it was dawn – with a horse and cart, myself and perhaps two other boys, and the father of one of the lads who had been lost, and we went down to the Battery, where the bodies had been laid out for identification. I remember they had tickets on them . . . Leurbost . . . Shawbost . . . Tolsta . . . and the man from Shawbost who went over with us, his son was there and I remember he was so handsome that I would have said he was not dead at all. I remember the colour in his face. I remember that fine yet . . .

His father went on his knees beside him and he began to take letters from his son's pockets, and there was money, I remember, silver and paper money, in the pocket of the trousers. And the father was reading a letter that he found and the tears were falling from him, splashing on the body of his son. I think it is the most heart-rending sight I have ever seen, and that was only one of many to be seen at the Battery that day – and for days afterwards.'

From all sides, ninety years on, the stories flood: the things that put detail and perspective on what might otherwise be one faraway hecatomb.

The remains of Kenneth Smith from Earshader were quickly found. The forty-four-year-old, a very strong swimmer, had been banged on the head, knocked unconscious: otherwise he might have made shore. His baggage was found, too, and those lovely silken gifts, all the way from Gibraltar.

They found, too, Kenneth MacPhail from Arnol, who had survived thirty-six hours afloat in the Mediterranean, the only survivor of his torpedoed ship. The *Iolaire* he had not survived and he was taken from the sea with both hands thrust deep into the pockets of his sea-coat, as if he had seen not the least point, now, in denying the ocean.

Angus Morrison, thirty-two, Petty Officer First Class, had waited overnight at Kyle for his wee brother, the Patch. They parted seconds before impact, as Angus scurried cheerfully away to collect his parcel. His body was never found.

Four officers and twenty-three crewmen had sailed the *Iolaire* to Kyle on 31 December. All four officers drowned and only the bodies of the engineers, Sub-Liuetenant Charles Rankin and 2nd Engineer John Hern, would be found. Rankin lies at Penzance; Hern in Sunderland. Of the twenty-three crew, only seven escaped. The casualties included Private Herbert William Head, of Ipswich, whose girlfriend was several months with child; their wedding was scheduled, in Stornoway, for New Year's Day.

Head's body was never recovered. And his lost *Iolaire* shipmates fill graves the length of the land – Havant and St Helens, Portsmouth and Newcastle, Southend and Auchterarder, Greenwich and Islington and Southampton. For six, from Mason to Herbert Head, the Portsmouth Naval Memorial is their monument, and there their names are cut with other unfound *Iolaire* men and all those Great War sailors whose only grave is the sea.

David MacDonald, the 'wee laddie frae Aberdeen', was drowned: seventeen and the youngest aboard. Leading Deckhand Charles Dewsbury, thirty-three, who had come from America of his own volition, lies miles and miles from home in the Sandwick cemetery, as does Assistant Steward Alfred S. Taylor.

Malcolm Matheson, 10 Upper Shader, on the West Side, had served as a gunner on HMT *Ireland*, which had felled two Zeppelins in the North Sea. Only ten of these German airships were brought down in the whole Great War. Matheson came home, only to die with the *Iolaire* and be buried at Barvas.

Some houses away, word reached Chirsty Morrison at 31 Upper Shader that her youngest son, Angus, was among the corpses lying at the Battery, a tarpaulin over his face and cold numbers chalked on his boots. His father was dead. His older brothers had emigrated to Canada. She was alone with her daughters, and Angus's young wife, Katy Ann. Mrs Morrison got her cart and her fat little pony and went over to Stornoway to bring home the remains of her twenty-year-old son. She ignored all well-meant offers of making the errand in her stead; she would hear of no one else going.

At 25 Lower Shader, the body of John MacDonald, thirty-three, trundled home. When his body was recovered, and his pockets emptied, an engagement ring had been found. No one ever knew who it was for.

Jessie Finlayson had married her man, John MacDonald of 10 Skigersta, only on 18 October. Joyously she had travelled to Stornoway to meet him early on New Year. But John had drowned. To add insult to devastation, the authorities declared she had no entitlement to a widow's pension, being childless. Her father exploded. How, he demanded, could the Admiralty be so sure there would be no issue of the marriage? The pension was hastily awarded, for they had no answer.

And so it went – the dragging, the hunting, the finds lying in this cove and floating in that bay, and the carts trudged back to townships, and little boats sailed from Stornoway to Lochs and to Harris with their honoured dead.

There was one ugly little scene in the harbour. The skipper of an East Coast drifter had been roped into the quest: having berthed, his crew began calmly hoisting up the remains with derrick-and-sling, dumping them on the quayside as if they were bolls of meal. He was threatened with physical violence, and wisely adopted more fitting means of bearing the fallen.

And it is said – Mary Ross could still remember it in 1972, in her Invergordon home – that Stornoway in a day or two ran out of coffins; and that more had to be sent for, and that a ship soon came from Kyle laden with them: and still, being found, and duly boxed, the boys of the *Iolaire* went home.

'I remember it as a six-and-a-half year old,' Calum Smith would write in 2001, 'who watched the horses and carts with their burdens of coffins going head to tail along the main road to the Bragar cemetery.' Hector MacIver remembered when the grim cortege came. 'That night, some beautiful Clydesdale horses lodged in our stable. They had come across the moor, it being New Year's Day, bearing an eldritch gift, the bodies of sailors who had been drowned in the disaster. I was summoned to come down and see the horses, for they were the first of their kind to be seen on the west coast. At first, afraid, I hid my face in the pillow.' His sister, born in 1912 and still spared in Shawbost in September 2008, can remember those carts going in the road, down Am Baile Staighe – though not quite as clearly, she says, as she can still see her much older brother Neil on 10 October 1916, his leave over, vaulting the garden wall to walk to Bragar and get a lift from Lakefield to Stornoway and the front, to fall in the last bloody weeks of the Battle of the Somme.

For four weeks – amidst first the influenza, the terrible Spanish 'flu now raging in the Western Isles as elsewhere, and then this blackest New Year – there had been no District News from North Tolsta. On 10 January, Duncan MacDonald flatly related that Corporal John MacLeod, 20 North Tolsta, had won the Croix de Guerre, on top of the Military Medal. A wedding on 26 December was described. At last, on 24 January, the catastrophe was mentioned – more accurately, a visit by Lord Leverhulme, occasioned by it: he called on 'the motherless children of the late John MacDonald, 1 North Tolsta', and 'met some of the local people who had applied for new holdings.'

Then, 'Since we have sent in our last report the remains of three of our local men have been recovered and buried in the local cemetery – Donald MacLeod and Malcolm MacLeod, sons of Malcolm MacLeod, 58 North Tolsta, and Evander Murray, 45 North Tolsta. As we were burying these at night, verses from "The Burial of John Moore" kept recurring to our memories with strange persistency. The bodies of Donald Campbell, No. 44, and of John MacIver, No. 69, have not yet been recovered.'

They never were. This midnight funeral, conducted by lamplight, typified the disintegration of normal funerary rites amidst an avalanche of death, and the exigencies forced, besides, by rapidly decaying bodies. But, a 'Cinematographer' had just visited the district to 'take photographs of the survivors and others. He also photographed sections of the township. We understand these photographs are to be exhibited in aid of the Disaster Fund.'

Three weeks passed. Peggy's uncle, Norman Morrison, farmer at Steinish on the outskirts of Stornoway, searched and searched and, after twenty days, found the body of her father, Seaman John Murray, by the rocks of Arnish. He did what had to be done and that body, too, was borne up to Ness and to Lionel. A single wake was held, and Peggy in January 2003 could still remember that ominous coffin, wrapped tight in roped tarpaulin. Murray was buried, like all the recovered men of Ness, in the old churchyard of Swainbost.

Six weeks passed. The body of Angus Matheson, a nineteen-year-old seaman from Uigen, had still not been found. One day his father, Calum, Càm – who had tried to speak with John MacLennan at New Year – rose early, and announced he was going to Stornoway to collect his son. He would give no explanation, but had evidently had a dream.

So his cart, too, made the slow, long journey by hill and moor, and when he reached the Naval Depot – it was late at night – he roused the officer of the watch and demanded a boat. They would go to Glumaig Bay and there, said Matheson flatly, they would find Angus's body. The officer was a shrewd, decent man. The night was good, calm; a boat was quickly organised, and they took Càm with them. And there, right in Glumaig Bay in the lea of Arnish Point, and just where the broken father directed, the grapples went down and up came the body of Angus Matheson. In 1978, a member of this little boat's crew could authenticate the story for Tormod Calum Dòmhnallach.

Months passed. In the summer of 1919, two very young men from Lochs were out fishing, and were badly shaken when a gruesomely putrefied body came in with the herring. But Angus MacDonald, twenty-three, from 3 Port of Ness, still wore his RNR tag. A surviving photograph of the young rating shows a most poised, handsome young man. There was no question of bringing these remains by cart all the way to the Butt, and he was buried quietly at Crossbost cemetery. His mother Margaret, Màiread Iain Bhàin, came to Crossbost Communion twice each year for the rest of her life, and the first thing she always did was to visit her Angus's grave. She died, aged eighty-six, in September 1952.

Yet, amidst all these things unspeakable, the sea – by the skin of its cold green teeth – was denied final victims.

A Harris women, resident in town, a Mrs MacKinnon, was frantic. Her man had not come. There was no word. She searched and searched and searched, and found a Naval cap – she turned it over and there, stencilled blackly, was the name JOHN MACKINNON. 'She went home to mourn the loss of her man,' Tormod Calum Dòmhnallach would write, 'And that afternoon a telegram came from her husband; he had remained in Kyle and was safe. Many in the islands share the same names.' Such, indeed, was the situation of the cap's owner: only two men called John MacKinnon, respectively of Tarbert and Ranish, had sailed on the *Iolaire* – and both lived.

But most remarkable was the experience of Chrissie MacKenzie, Cairstìona Dhòmhnaill 'Ain Bhàin, of Post Office Side, Cross, in Ness. Word had come from town – Norman, her eighteen-year-old brother, was among the dead – and, her parents being utterly distraught, she made the journey over the moors to identify him. And there he was indeed, on the floor of the ammunition store, soaking, covered, numbers on his boots. She tightened her jaw, managed a nod, and they lifted the tarpaulin, and there she beheld the body of her little brother. She nearly broke down then, on the spot, but Chrissie knelt, and felt the icy hand, and looked in full tide of emotion at the slack white face.

And there was the tiniest flicker in the eyelid, which only she saw.

Chrissie MacKenzie hit the ceiling. The officers turned as pale as the supposed corpse. John Mackenzie was bundled off to Lewis Hospital and, after days of anxiety, and long convalescence from assorted injuries, made an entire recovery, and my father remembers Làrag in Cross, a well-known personality, driving his bus, getting on with his life. His was a classic case of 'near drowning', of which few doctors in 1919 would have known: a powerful, mammalian diving-reflex where in certain dire conditions the body kicks powerfully into suspended animation, heartbeat slowing to as few as six lup-dups a minute, concentrating blood to the lungs, brain and heart itself. The younger the victim, the more active the reflex; the colder the water, the higher the chance of it, and an essential trigger seems to be entire facial immersion. Resuscitation should always be attempted and there are well documented cases of children and teenagers surviving, and restored to full health and vitality, after even thirty minutes completely under chill water. That New Year, might other boys have been revived?

Angus MacDonald's of Port of Ness seems to have been the last of the *Iolaire* burials. Fifty-six of the 205 dead would never be found; eight more, duly landed, were both tagless and beyond identification. The men of the *Iolaire* dot almost every cemetery on the Long Island. They huddle by the ruins of the tiny medieval church in the Bragar cemetery. They lie at Bosta, Gravir, Swainbost, Barvas, Valtos, Ardroil, Dalmore, Laxay, Luskentyre, Sandwick and Gress. They sleep by Cladh Mhìcheil in Tolsta, in the ranked rows of Aignish and in the machair of Berneray, and they lie in family plots with those of their own blood, and not together, but laid no doubt where they would have chosen to be.

They are almost all kept in remembrance under small, uniform granite headstones of the First World War, each with the crest of the Royal Naval Reserve – an anchor ensnaked by rope – and in most instances giving no more detail than a name, a service-number, a ship – as likely to be that of last service as the wretched steam-yacht – and the bleak date, '1st January 1919'. Some, at relatives' insistence, had their age appended, or a line of Gaelic scripture, or a brave endearment.

Most powerful, perhaps, is the ranked mass-grave at Sandwick, where eight identical headstones attest only

A

SAILOR

OF THE

GREAT WAR

ROYAL NAVY

HM YACHT "IOLAIRE"

1ST JANUARY 1919

KNOWN UNTO GOD

Two hundred and eighty-four men had sailed from Kyle of Lochalsh at 7.30 p.m. on New Year's Eve. And 205 were dead. Of the sixteen Tolsta passengers, only five lived. Leurbost in Lochs, too, had lost eleven men. Nine of the Shawbost passengers were dead. Tiny Crowlista, in Uig of the mountains and the beaches, had entrusted six men to the *Iolaire*, and all six had died: Crowlista has never recovered.

And Sheshader, in Point, had sent ten men aboard, and not one came off alive, so that it was said 'Chaidh Seiseadar a bhàthadh,' 'Sheshader was drowned.' Thirty-seven-year-old Norman Montgomery, 3 Sheshader, left a widow and six young children. As a boy of eight, little Norman had fallen over a nasty local cliff onto the beach. No one saw. His cries were heard after dusk by a blind man, Donald MacKay, who knew every rock and turn in the district by smell and by hand. MacKay made unerringly for a gully, climbed down the precipice, retrieved the badly bruised child, climbed back up with the mite on his back, and bore him home to his panic-stricken parents. For this Norman Montgomery had been marvelously delivered: to die with the *Iolaire*. His body was never found.

There are still worse stories, as Chris Lawson remembers from her late mother, Mary Anne MacDonald, 'how she was over with Doileag, Dolina MacLeod at 15 Sheshader, one day. Doileag, who was her cousin and had grown up in the house next to them on croft 11, had written a letter to her husband. They were standing outside the house when two village lads, brothers at home on leave, went along the road. They were heading for the Post Office in Aird, and Doileag asked them to post the letter for her. The two lads were the two John MacDonalds – brothers at No. 20. They and Doileag's husband were all lost on the *Iolaire* months later. That image was etched on her mind all her life. She was fourteen at the time . . .

She also remembered Doileag's sister coming in that day and Doileag's children with her. She made the children go outside which was unusual until she related to them what had happened with the *Iolaire* and their father. Nancy MacKenzie, one of the children, only died a few years ago and she told me shortly before she died of how her grandfather had come in to the house after church and made her mother, Doileag, sit down. He told her then of the tragedy. Her only memory of her father was of a big man in uniform carrying one of the others on his shoulder when he was home on leave. Her youngest sister was Jessie MacKenzie, the well-known Gaelic singer, who was a babe in arms when the father was lost. Nancy also told me that the following New Year she and a friend went to William Murray's mother's house at No. 11 to wish them a Happy New Year as they used to do. William Murray's mother, who was my mother's aunt, told them to leave and they could not understand why she did not let them in . . .

And this was only Sheshader.

Three men lost besides in Aird, two in Portnaguran, five in Shulishader, four in Upper Bayble, one at New Park in Bayble, one from Broker, two from Garrabost, three from Swordale, three from Knock, two from Aignish . . .

And this was only Point.

Twenty-three men would never return to the townships of Ness. One, forty-three-year-old Roderick Morrison of 89 Cross Skigersta Road, left his Margaret a shattered widow, now alone to bring up their eight young children. Eight sons, brothers, husbands of Bragar were dead. Ranish in Lochs lost five men; little Lemreway lost three.

And all the other threes, and the fours, and the ones, and the twos, were no less rending for all bereaved and afflicted; such were no less bewildered to lose their menfolk not by mine or by torpedo, neither in the mud of Flanders nor in the skies over Belgium nor the bloody surf of Gallipoli, not by sword or bullets or otherwise by the sound of trumpets, but drowned incomprehensibly, pointlessly, amidst peace renewed and at the very gates of home.

And an enduring memory of the small boy, Donald Morrison, brings near the most intimate, harrowing pain of the tragedy. 'A couple of days after, when my mother got out of bed and she was able to go around the house, she started opening kit-bags, and a case he had, and he had something for every one of us – a wee present for the New Year, or for Christmas. And I remember – I can remember my mother . . . it brings back memories . . . when she was going through all this stuff – oh, she was crying, you know – 'course, we started crying with her, though we were young at the time . . .'

'The remainder of the unfortunate victims,' said the *Stornoway Gazette*, in sweeping dismissal, 'were natives of Harris, or members of the crew.' But the men of the ship's company were no less precious to those who loved them, and of the eleven Harrismen who had snuck aboard the *Iolaire*, only four had been spared. It would take the *Gazette* seventy years to print the names of the crew; over eighty before it at last honoured the Harrismen.

And grief, wrath and incredulity now combined, demanding account for so dreadful a waste of life.

Tired, overwhelmed, Rear-Admiral R.F. Boyle fumbled with his notes and tried to gather his thoughts. At least 200 men were dead and, to complicate his life still further, so were all the officers of the *Iolaire* . . . and already, though it was not yet twenty-four hours since the *Iolaire* had sailed from Kyle, he knew that rage was rapidly mounting in Stornoway and beyond. Within a day or two the atmosphere might be dangerous, and Rear-Admiral R.F. Boyle worried for his own back.

He had been slow to comprehend the scale of the catastrophe, as an uncomfortable telegram late on New Year evening attests.

(0530) Sent 10.45 pm 1st
Received 12.15 am 2nd.
Telegram 601
TWO FURTHER SURVIVORS HAVE REPORTED ALEXANDER MACKENZIE
DECKHAND HMS VALOROUS ADDRESS 10 LUERBOST LOCKS [SIC]
AND ANGUS NICHOLSON [SIC] LEADING SIGNALMAN RNR 5136B HMS
IMPERIEUSE ADDRESS 10 BATTERY PARK STORNOWAY. IOLAIRE TOTAL
WRECK AT HOLM POINT NORTH TRUE OF BEASTON HOLM [SIC]. TO
PRESENT ONLY 60 SURVIVORS KNOWN. MAY POSSIBLY BE MORE THAT
HAVE DISPERSED INTO COUNTRY. IT WILL TAKE SOME TIME TO FIND
OUT NAMES OF THOSE WHO ARE LOST AS THERE WAS NO LIST TAKEN
BEFORE EMBARKING.
(2225)

He was a pen-pusher, really, this Rear-Admiral Boyle, but shrewd; he had himself little doubt where immediate blame for the disaster belonged.

Chapter Six
'Searching and Impartial Inquiry'

'Reports that say that something hasn't happened are always interesting to me, because as we know, there are known knowns; there are things we know we know. We also know there are known unknowns; that is to say, we know there are some things we do not know. But there are also unknown unknowns – the ones we don't know we don't know.

If I said yes, that would then suggest that that might be the only place where it might be done which would not be accurate, necessarily accurate. It might also not be inaccurate, but I'm disinclined to mislead anyone.

There's another way to phrase that and that is that the absence of evidence is not the evidence of absence. It is basically saying the same thing in a different way. Simply because you do not have evidence that something does exist does not mean that you have evidence that it doesn't exist.'

Donald Rumsfeld, US Secretary of Defense, 2002

'They are hardy, intrepid, accustomed to a rough country, and no great mischief if they fall.'

General James Wolfe (1727–59) aide de camp to General Henry Hawley at the battle of Culloden; commander at the siege of Quebec – on Highland soldiers

On Thursday, 3 January 1919, the Stornoway fathers convened 'in terms of a requisition by the Provost to consider as to the Town Council taking steps to open up a subscription list or funds in aid of those who are dependents of the sailors who lost their lives in the wreck of HMY *Iolaire*.'

Top of the agenda was a message both of condolence and demand.

That this Town Council deeply deplore the appalling disaster which happened to HMY *Iolaire* at the entrance to Stornoway Harbour on the 1st of January instant, by which over 200 lives were lost under the most tragic circumstances, and express on their own behalf and in the name of the community, their profound sympathy with the dependents and relatives of our brave sailors who perished on the threshold of their homes as they were looking forward to a happy reunion with their loved ones after a prolonged period of war service, and they demand of the responsible authority the strictest investigation into all the circumstances attending the catastrophe and the responsibility attaching thereto.

The Council had to acknowledge untold telegrams of sympathy, from the highest in the land – 'Their Majesties The King and Queen were shocked to hear of the disaster . . .'; Alexandra the Queen Mother ('more distressed than I can say at this unspeakably sad disaster which has befallen our dear sailors at Stornoway. Please convey my utmost and deepest sympathy to relations at the heartrending calamity which has deprived them of those nearest and dearest to them at the moment of their return home after their splendid services in the past four years') – through the Secretary for Scotland, Robert Munro; three different messages, under three of his different hats, from the indefatigable Lord Leverhulme; Sir Henry Morgan; Sir Robert Leicester Harmsworth, the MP for Caithness and Sutherland; Wick Town Council; assorted military officials; several ministers; several schoolmasters; the National Bank of Scotland; E. Woodger and Son, Lowestoft; and Mrs Craven, Whitney.

The Town Clerk then pondered the needs of families, and decided to liaise with the Lewis District Committee – the local authority, effectively, for the 'landward' part of the island and the villages largely hit by the calamity – and in due time a meeting was fixed for 13 January.

For all our lives we have been gently lullabied into perceptions of the Highlander as victim – fey, gentle, passive little things, padding obediently after Stuart princelings or trudging onto an emigrant ship to a chorus of Gaelic song and the odd 'Ochone'. But the people of Lewis, this New Year of 1919, neither wallowed in self-pity nor quailed before high authority. By nightfall on 1 January, the island sought answers. By the Thursday evening, Lewis had not only launched plans to guarantee at least minimal financial security for the grieving families, but now lobbied with vigour for ruthless inquest. When the Admiralty denied them, they successfully moved distant Government to intervene and – what was more – forced the Royal Navy to cooperate, in a humiliating climb-down.

It was all the more remarkable, as it was only in 1911 that it been established the Crown could be held liable in civil law. The principle was laid down in the case of George Archer-Shee, a young cadet falsely accused of theft at Osborne Naval College, who won subsequent vindication in court and compensation from the Admiralty. Indeed, much justice was won for the *Iolaire* families without any resort to litigation. The Fatal Accident Inquiry of February 1919 could not provide definitive answers, but it did ascertain definitive facts and ended with a broadly fair indictment – and all this against the dismal parade of *Iolaire* bodies and raging influenza, on an island reeling from wholesale death and in a town that had just run out of coffins.

Driving all this was one insistent rumour: that the officers of the *Iolaire* had not been sober. 'This demand of a full investigation barely concealed powerful local suspicions about the fitness of the ship's officers on the night in question,' writes Tormod Calum Dòmhnallach. 'New Year's Eve in the islands is traditionally a time for celebration, when much whisky is consumed.' 'Statements of survivors as to the primary cause of the disaster, and the allegations made by them, are causing deep unrest in the minds of the bereaved,' darkly declared *The Scotsman* on Monday, 6 January 1919.

For many years temperance had been a huge issue in Scotland, especially on the rising forces of the Left and in the nascent Labour Party – and not least on Lewis. In 1884, a Stornoway minister, Rev. Donald J. Martin, had actually bought up a pub, the Star Inn – the oldest building, then and now, in the town, where men had plotted for the 1719 Jacobite Rising – and turned it into a coffee-house. The Star Inn has long since been restored to the service of Satan; but in 1919, Prohibitionists were a powerful body of island opinion and enjoyed particular support in the United Free Church. She already experimented with non-alcoholic Communion wine and agonised if the recovered drunkard might legitimately take the sacrament with only the bread. Overseas, the Volstead Act would shortly launch America on its disastrous experiment with Prohibition (thousands would die from home-brewed, toxic hooch; organised crime would gain a grip it has never since relinquished). At home, Stornoway herself was on the brink of long 'dry' years.

It was natural that distraught *Iolaire* families would immediately suspect John Barleycorn and, within forty-eight hours of the wreck, the wildest tales sped all over Lewis. They had one obvious pretext – the collision of the *Iolaire* with Kyle pier on the 31st, witnessed by many Lewismen.

But the pressure on the Admiralty was increased further by a new political fact. Since the late general election, in December 1918, the Western Isles now had its own MP, the earnest Dr Donald Murray, hitherto a dedicated Medical Officer of Health. In alliance with the national press – most papers quickly sent a hack to Stornoway – Dr Murray lobbied hard for public investigation.

Boyle privately knew that the disaster was, indubitably, the fruit of gross incompetence – a near-incomprehensible blunder in navigation. His immediate instinct was to arraign the dead officers. On Friday, 3 January, he telegraphed the Admiralty: 'Request instructions as to whether court martial should be held on the loss of yacht *Iolaire*.' Meanwhile, he sat down to dictate a formal statement for his superiors. It would be hidden from the people of Lewis, with all other Admiralty papers on the disaster, until File 693 was opened to scrutiny in January 1970.

Sir,
I have the honour to forward the following report of the sinking of HMY *Iolaire* on 1st January 1919.

Iolaire had been sent to Kyle on 30th [sic] December 1918 to augment the regular steamer service in conveying ratings to Stornoway which were expected to arrive at Kyle on 31st December 1918.

The number of ratings that arrived in Kyle on 31st December was more than double the number which had been notified. *Iolaire* left Kyle at 7.30 p.m. and it was reported from Kyle that 260 ratings were on board.

At 1.55 a.m. it was reported that a ship required a pilot. At 2.20 it was reported that vessel was sending rockets – at this time the wind had freshened to a southerly gale, the atmosphere remaining clear.

A drifter was sent out with Lieutenant Wenlock in charge. They failed to locate the wreck.

The Secretary of the Life Boat was called on to launch the life boat but he and the Coxswain were unable to get a crew. It is doubtful if the lifeboat could have rendered any service.

The only available vessels, one drifter and the HM Whaler *Rorqual* who had returned from dock on the 31st were sent out to stand by.

At 3.30 a.m. 5 ratings from *Iolaire* reported at my office. Little information could be obtained from them; they appeared to have been washed on shore or else dropped from the stern when the vessel's stern was swung over the rocky shore at Holm Point – in fact they did not know exactly how they got on shore.

At daylight nothing could be seen of *Iolaire* except her masts – the foremast had carried away just above the hounds, the Mainmast was standing and one rating Donald Macdonald [sic] was taken off having remained aloft when the *Iolaire* sank – this rating apparently had not suffered from his exposure.

Search parties were sent out at daylight and during the day 18 bodies were recovered. The work of recovering bodies continues with the aid of volunteer fishermen who are dragging.

Further evidence is being taken but as yet there is little information as to how the disaster occurred.

The *Iolaire* is in the position reported by my telegram timed 20.40 2nd January 1919 viz:- Lat. 58.11'18" N., Long. 6.21'30" and is probably a constructive total wreck.

I have the honour to be, Sir,
Your obedient servant,
R F BOYLE
Rear-Admiral

The *Iolaire* had, of course, sailed from Stornoway on the 31st. The report of a ship requiring a pilot may have turned on misunderstanding of one word – a 'blue light' had been seen, but in 1919, as one or two statements in File 693 unconsciously attest, a rocket (and especially a non-explosive rocket) was often described as a 'light'. A blue lantern, though, hung on the bridge, was the usual signal for a pilot; the blue rockets had been a cry for help. The misunderstanding was not Boyle's, who got the report third-hand, but Ainsdale's – it nevertheless lost precious time. Wenlock actually had located the wreck – though he could not approach – and it is most unlikely exhausted *Iolaire* survivors were in Boyle's office as early as 3.30 a.m. We might note there is here no reference to the involvement of Wenlock or any other RNR officer in the retrieval of Donald Morrison.

On Thursday 2nd, Royal Naval divers from the Stornoway base, after laboriously donning the cumbersome kit of the day, descended to the wreck. Thomas George Gusterson, from Sussex, was stalked for the rest of his life by what he had seen – 'I found myself among a sea of bodies' – and flatly refused to go down again. Later that night, the Rear-Admiral could telegraph details of the wrecked hulk. 'Divers report yacht *Iolaire* completely broken in two abaft foremast. After part completely wrecked except skin of ship. Divers think boiler blew up but this is not borne out by evidence of survivors and mainmast is still standing.' This, too, is interesting. Any iron steamship will make loudest report when her back breaks; and with her furnaces still ablaze with coal one might well expect much flash and hiss in the instant of disintegration. The 'explosion' at foundering became a very early part of enduring *Iolaire* folklore and, even by the time of the Fatal Accident Inquiry, vivid press reports had evidently convinced some survivors they must have seen one.

Within twenty-four hours – at 12.30 p.m. on Saturday the 4th – there was a curt reply to the request for a court martial. It was refused. 'Court of Enquiry is to be held.' Boyle's heart sank. A Court of Inquiry (to use the Scots spelling) required far more work; could be of broadest scope, focusing pitilessly on the chaos of rescue efforts from land; and might even implicate him. His mood cannot have improved when he received formal intimation that day from Archibald Munro, Town Clerk, of the Town Council resolution. It was forwarded to London and, on 18 January – less than candidly; Admiralty inquiries were by then complete – their Lordships were pleased to assure Stornoway Town Council that the 'cause of the disaster will be fully investigated.'

Boyle busied himself with details, even telegraphing on the Sabbath with fussy reflections on the wreck at Holm. 'Divers report is unreliable. By using waterglass at LWS [low water spring-tide] it appears that the sheer deck aft is intact. Back is broken as previously reported. There is no indication that boiler exploded. Bridge and engine room skylight glass appear to be undamaged.'

But still more important – and sent deliberately in 'cypher', code – was another Fourth Commandment-busting exchange with Commander Walsh at Kyle, as the talk on Lewis grew wilder yet.

From: Rear-Admiral, Stornoway
To: Port, Kyle
Date: 5th January, 1919
RUMOURS HAVE REACHED ME THAT THE OFFICERS IN *IOLAIRE* ON 31ST
DECEMBER WERE NOT IN A FIT STATE TO TAKE CHARGE. REQUEST YOUR
OPINION.
1315

From: Port, Kyle
To: Rear-Admiral, Stornoway
Date: 5th January, 1919
YOUR 1314. FROM PERSONAL KNOWLEDGE AND FROM INFORMATION
RECEIVED CAN STATE COMMANDING OFFICER AND NAVIGATING
OFFICER FIT STATE TO TAKE CHARGE. HAVE NO REASON TO SUPPOSE
OTHER OFFICERS WERE NOT EQUALLY FIT.
1516

Boyle knuckled down to his Court of Inquiry, and on 7 January appointed three officers by letter to 'assemble at 10 am on Wednesday 8th January, as a Court of Enquiry; and hold a full and careful investigation into the circumstances attending the stranding and loss of HM Yacht 056 *Iolaire* on Biastan Holm Rocks about 0220 on Wednesday, 1st January . . .' The 'report is to be accompanied by the Minutes of Evidence taken, all in triplicate, and is to state whether blame is attributable to anyone in this matter, and, if so, to whom, and to what extent, and is also to state the steps taken to save the officers and men on board.' The Rear-Admiral had ordered all survivors to report to Stornoway 'today' (7 January) for the purpose of the Inquiry. And he called attention to Articles 703 and 690 of the King's Regulations.

The wording is sly: explicitly judicial – to assess blame, rather than to ascertain facts – and directed attention on the 'stranding and loss', rather than the less than impressive efforts at rescue from land: these to be stated, not assessed. Boyle was, besides, mistaken – or out of date – on the King's Regulations, as a tart legal hand later noted in File 693. He may have been working from an obsolete edition with the Articles differently numbered; he undoubtedly meant those Articles relating to production in evidence of 'a ship's log books, table of compass deviations, order books and work books and the charts and sailing directions by which the ship was navigated', and as every last sheet was now sunken mush by the Beasts of Holm, his legalese was irrelevant.

Boyle demanded statements from everyone he could think of – Commander Walsh; Lieutenant Wenlock of the *Budding Rose*; Sub-Lieutenant C.W. Murray; and Boxall and Barnes of the Coastguard. And he granted survivors little time to recover. The letter he would finally send to London on Tuesday, 14 January 1919, accompanying the 'Minutes and Findings of the Court of Enquiry', attests to his obsession with detail and easy disregard for traumatised sailors, at least when Boyle had forms to fill, boxes to tick.

Before the Court was held the following arrangements were made for collecting the evidence of the survivors from HM Yacht *Iolaire*.

. (1) Early evidence was taken from those who were in the Sick Quarters and those of the *Iolaire*'s crew who were survivors.

(2) All survivors were informed that their presence was required at the Base on Tuesday 7th January, with a view to weeding out those whose evidence was of no value. Each man was interrogated and those required for the Court were informed that they were required on 8th January to attend a Court of Enquiry. These amounted to 25. Those not required were allowed to return to their homes in the various parts of the Islands. Some men were prevented from attending on medical grounds.

From the evidence there appears to be nothing to account for the disaster. None of those on watch at the Bridge at the time are survivors.

Ill-favoured rumours were circulated concerning the condition of the officers on board. These appear to have risen from the fact that while berthing at the East side of Kyle Pier the *Iolaire* collided with the pier. She was turning to berth with her head to the southward – a difficult manoeuvre at any time – sufficient allowance was not made for the west-going tide and the *Iolaire* had rather much way on. This was considered an error of judgement, and in no way through carelessness.

The Master of the Mail Steamer *Sheila* (which runs between Kyle and Stornoway) witnessed the above, and the above is his opinion of the occurrence. He also informed me that he saw the *Iolaire* leave the pier, after embarking the libertymen, and considered she was handled in a most seamanlike manner.

A report from Lieutenant Commander Walsh relative to the proceedings at Kyle is attached.

Feeling still running high, on the 5th January I sent a cypher message to Lieutenant Commander Walsh, copy of which with his reply is also attached.

The evidence goes to show that the rumour is quite unfounded.

The Commanding Officer, Commander Mason, RNR, appears from the evidence to have gone below shortly after 1 am on the 1st January, and Lieutenant Cotter, RNR, appears to have been in charge of the ship.

The large number of ratings arriving at Kyle was quite unexpected. The *Iolaire* had made two similar trips with libertymen during the first leave, with Lieutenant Forcett H. Skinner, RNR, in charge and with a half crew. In view of the fact that the ratings were looking forward to spending the New York [sic] at home with such eagerness, it was natural for Commander Mason to take such a large number, although there were not sufficient boats or lifebelts to meet a case of urgency.

As far as can be ascertained the number of persons on board *Iolaire* at the time of accident was 284. (24 of the crew and 260 passengers.) Of these the number of known survivors is 79; the number of bodies recovered up to the present is 117, of which all have been identified except 9.

Copies of reports from the undermentioned officers are attached:-
Lieutenant Commander A.H. Walsh, RN, Kyle
Lieutenant A.W. Wenlock, RNR, Stornoway Depot
Sub Lieutenant C.W. Murray, RNVR, Stornoway Depot
Mr F. Boxall, Divisional Chief Officer, Coastguard
Mr W. Barnes, Acting Chief Officer, Coastguard

I have the honour to be, Sir,
Your obedient servant,
R F Boyle
Rear-Admiral.

A complete copy had been forwarded to the Commander in Chief, Coast of Scotland, Rosyth.

The survivors were imperiously summoned to Stornoway by telegrams, sent to every village post office by 7 January. (That received by Alick Campbell of Plocrapool survives – his nephew, Alasdair MacNeil, proudly showed it to me in January 1999 – and seems most rude to our modern sensibilities: an explicit order. The young Harrisman had, with scarcely any respite, to retrace his steps to Stornoway and, like everyone else, endure cold interrogation with not the least effort to make the experience easier.)

Yet none of this evidence taken in initial questioning survives, an unfortunate and perhaps suspicious omission in File 693. All the survivors were tired, sore, trying to come to terms with a dreadful experience. It was no easy matter to refuse on health grounds. John MacLennan in Kneep, his legs and chest badly cut, somehow managed to evade it. In Tarbert, Dr K.C. Crosbie was exasperated when the Admiralty demanded John MacKinnon's presence. 'I certify that I have today examined Seaman John MacKinnon and find him suffering from catarrh and a bad cough. Also pains in side. In my opinion he is unfit to travel,' he penned acidly on Wednesday, 8 January. MacKinnon was grudgingly excused, but a sworn statement survives in the file.

Everyone else made for Stornoway – some by boat, some by the few rattling motor-vehicles, some by cart, most by foot – and were duly 'weeded out', as Boyle graciously put it. Twenty-five had to stay overnight for a minuted grilling at the Court of Inquiry. These were not chatty proceedings; and all the more fraught since, athough these island ratings had good plain English, it was neither their first tongue nor natural language of thought.

And, along with their eagerness to vaunt the sobriety of officers, Boyle's policemen were almost as obsessed with the lifeline rope. They evidently hoped to prove Mason or Cotter had ordered it, and there was some confusion from witness testimony as to who, precisely, had made the death-defying swim. Few had witnessed MacLeod's dive and, in dark and storm, could have seen scant detail.

'My name was the first that was called,' Donald Morrison recalled ruefully in 1968. 'I was called in and they asked me if I knew the man that went ashore with the rope. I told them that he was here, that he was outside. So they asked me to go and fetch him.' John Finlay MacLeod, fifty years later, could not hide his dry amusement. 'They seemed to get somewhat mixed up in it, somehow. Suspicion about that rope – they got it mixed up with this line here. A few said they pulled the rope ashore. But that's a different story all together . . .'

In the event, twenty-four passengers and crew were examined on 8 January, most after preparing a sworn statement. James MacDonald, skipper of the fishing boat *Spider*, also put in written testimony, but does not appear to have been cross-examined and, as a civilian, could not have been so compelled.

A vital witness was first to testify:

James MacLean, no 28 Shore Street, Campbeltown

I am a fisherman, but in July, 1915, I joined the Navy for the Patrol Service, and on 8th February, 1916, I was appointed as a deckhand on the Yacht *Iolaire*.

I was on board said Yacht on the morning of 1st January, 1919, when she was lost on the rocks near Stornoway. We were proceeding from Kyle of Lochalsh to Stornoway. The morning was dark but there was no fog and lights could be seen quite distinctly. I was at the wheel from 12 to 1 on said morning. I was relieved by deck-hand Ernest Leggett, and after leaving the wheel went to the bridge and promenade deck to be at hand for any duty. When I was at the wheel Captain Mason was in command and at 1 am Lieut. Cotter took over charge. He was the Navigating Officer. When on the bridge leaving the wheel I reported the lights of Stornoway to Lieut. Cotter, who said 'All right.' At 1.30 or about that time, he sent me to the main deck to call the hands to be ready to let go the anchor. I did so, and returned to the promenade deck. Lieut. Cotter told me to go down again to see that the hands were up and everything was ready to let go the anchor. I was just returning to the promenade deck when the vessel struck the rocks. I had had no idea that the vessel was off her course. She took a terrible list and in about an hour's time sank. I was saved by getting hold of a rope and managed to scramble on to the rocks.

Leggett who relieved me at the wheel was perfectly sober and so was the Navigating Officer. I cannot account for the vessel having been off her course, but there was no reason why she should have been.

The Accident happened about 5 minutes to 2.

There was no one on the bridge except Lieutenant Cotter and Leggett and myself as before mentioned. I was not on the bridge when the accident happened.

All which is truth.

SAYS FURTHER
There was a strong breeze from the South when the accident happened. It afterwards increased to a gale.

Truth.

(signed) James MacLean

An ambiguity in the first paragraph should be noted: had MacLean on 6 February 1916 joined the original *Iolaire* – which had only left Stornoway after the Armistice – or the *Amalthaea*, the ship that sank on 1st January 1919? Though probably the latter, it is nowhere clarified. It is also evident the wheelhouse was not actually on the bridge – this is not as ridiculous as it might sound, as a visit to the Royal Yacht *Britannia* at her Leith moorings will confirm. Though built in 1953, her wheelhouse is below the bridge and with no external view whatsoever – the helmsman stared straight into a bulkhead, steering per instruction from above. MacLean evidently had some outlook to sea – under fierce questioning he expressly confirmed he could see the light on Arnish Point when he was steering (or at least a lighthouse somewhere), but it may have been most restricted.

Another fact should be borne in mind. The bridge and wheelhouse of the *Iolaire* were a long way from her bows – 'more than half her length', Alexander Reid observed in his technobabble postscript to Dòmhnallach's *Call na h-Iolaire* in 1978, 'or 100 feet from her bowsprit, due to her yacht lines – a time-lag of six seconds at ten knots.' In other words, a look-out at the bow would see any light or hazard six seconds before anyone on the bridge. But this half-crewed ship had no such sentinel on the fateful night, any more than she had adequate life-saving equipment, or a radio that actually worked. And radar, bridge-controlled engines and variable-pitch propellors (whereby a ship can be put quickly astern without stalling and reversing her machinery) were still decades away.

As an instance of how the Court treated witnesses, we may note the questioning of the Port of Ness hero, John Finlay MacLeod.

389. Q. Where did you go?

A. I went aft.

390. Q. How did you land?

A. When she got close on to the rock, I jumped right over the stern with the life-line in my left hand. That is how I got ashore, between swimming and being tossed about.

391. Q. Are you the only man who landed at that moment?

A. No, there was nobody with me.

392. Q. What did you do when you landed?

A. As soon as I got up I was caught by the sea and I hung on until it got slack again. Then I got off. I saw a man on the top of the rock, and I called to him.

393. Q. Had you got the line in your hand then?

A. Yes.

394. Q. Did the man come to your assistance?

A. Yes, we stretched the line as taut as we could and they started coming out on the line then.

395. Q. What size line was this?

A. A heaving line.

396. Q. How many men came ashore by that line?

A. I should say four or five by the heaving line.

397. Q. After that, what happened?

A. We told them to get a heavy rope on the line. We pulled this ashore and we got this round the rock and somebody was hanging onto the end of it.

398. Q. How long was it kept round the rock?

A. According to my estimate, I should say about half an hour. It was there all the time until I left, but it was shifted to midships.

399. Q. What made you leave the rope?

A. I was exhausted. I could not hold on – it was impossible. The cold was too much for me.

Further comment is superfluous. In all, twenty-five witnesses were heard – James MacLean the deckhand and AB; Leonard Welch the telegraphist; Arthur Topham the trimmer; Ernest Adams and Griffith Ramsey; James Wilder, deckhand; and John McLellan or MacLennan – all the surviving *Iolaire* crew; and nineteen passengers – Archibald Ross (Leurbost); John F. MacLeod (Port of Ness); Murdo MacFarlane (Cross); Donald MacIver (North Tolsta); Donald MacRae (Ranish); John Montgomery (Ranish); Alexander or Alasdair J. MacLeod (Coll); Murdo MacDonald (North Tolsta); John MacInnes (North Tolsta); Kenneth MacLeod (Swordale); Angus MacDonald (Bayhead, Stornoway); Alex MacIver (Church Street, Stornoway); Angus Nicolson (Battery Park, Stornoway); Malcolm MacRitchie (Aird Uig); John MacKinnon (Ranish) – and the unsinkable Neil Nicolson, a 21-year-old rating from Lemreway who, called to the colours in 1916, had survived the sinking of HMS *Avenger*, the foundering of HMS *Campaign*, had just escaped the disaster of the *Iolaire* and would finally outlive every other man in this story.

By the close of that same day – the speed is more disconcerting than impressive – Rear-Admiral Boyle had his verdict.

COURT OF INQUIRY REPORT

HM Naval Base
Stornoway
8th January 1919

Sir,

We have the honour to report that in compliance with your memorandum No. 19/13 of 7th January, we have held a strict and careful inquiry into the circumstances attending the loss of HM Yacht 056 *Iolaire* stranded on the Biastan Holm Rocks at approximately 0150 on Wednesday the 1st January, and then drifted to the shore to the northward of the rocks and became a total wreck.

There is no evidence to explain how the accident occurred as none of the officers on board, or the helmsman or lookouts who were on deck at the time are among the survivors, and no opinion can be given as to whether blame is attributable to anyone in the matter.

The Court is further of opinion that no adequate or properly organised attempt was made to save the lives of those on board; the only steps taken were to fire rockets, burn blue lights and blow the whistle to attract attention. The boats appear to have been lowered without orders or guidance, and a hawser was got ashore from the stern. Orders from the bridge were subsequently given to bring the hawser amidships to gain a better lea.

With reference to Article 690 in King's Regulations, it has not been practicable to attend to the points mentioned therein owing to no records being available.

We have the honour to be, Sir,
Your Obedient Servants,
J.G. Humphreys, Commander
W. Bradley   "   RNR
W.A. Westgarth, Lieutenant RNR

This Jesuitical prose deserves comment. One helmsman on deck at the time – James MacLean – was certainly among the survivors, if not actually at the wheel when the *Iolaire* smashed into the Beasts. The Court evidently could not bring itself to admit that a mere rating had of his own initiative got the vital hawser ashore; and there is no evidence at all that the hawser was shifted amidships on any order from the bridge. On wider reflection, there seem abundant grounds to blame both Commander Mason and Lieutenant Cotter; the Court had certainly taken Boyle's strong hints and offered no judgement whatever on lifesaving endeavours from the shore.

The Rear-Admiral still fretted over the aspersions on officers. On 12 January, he sent an insistent note to Commander J.G. Humphreys, now grandly styled 'The President of the *Iolaire* Court of Inquiry'. The reply was on the morrow and, as far as Boyle and his appointed inquisitors were concerned, final.

In view of the rumours which are afloat in the neighbourhood that the Officers of HM Yacht *Iolaire* were not in a fit state for duty on the night of the 31st December, you are to arrange as far as possible for all the witnesses who saw, or spoke with, any of the Officers, to be re-examined as to their condition.

R F Boyle
Rear-Admiral

Sir,

We have the honour to report that in compliance with your Memorandum No. 19/13 of 12th January, we have this day re-examined eight witnesses who either spoke with, or saw, the Officers of HM Yacht *Iolaire* on the night of the 31st December 1918, and are of the opinion that the rumours that these officers were not in a fit state to perform their duties on that night are untrue and entirely without foundation.

The ink was barely dry on the Court's curious Report when everybody who was anybody on Lewis called for its findings and verdict to be made public. Boyle refused. On 14 January, he posted the weighty parcel to the Admiralty. The calls became a sensation. And another entirely self-inflicted blow embarrassed the Rear-Admiral. Within a fortnight of the disaster, and with eighty-eight bodies still missing, Rear-Admiral Boyle had agreed the sale of the wrecked, sunken *Iolaire* to a mysterious Mr White. By the 15th, as an alarmed exchange of telegrams between London and Stornoway makes plain, this was public knowledge. Now there was uproar.

'It is hard to see why the Admiralty was so obtuse,' observes Dòmhnallach. 'The local demand for a Public Inquiry became a clamour.' At Boyle's belated beseeching, the Admiralty liaised with Mr White on the 15th and assured the Stornoway commander that all plans for sale and salvage were suspended and that the scrap-merchant 'is quite willing to withhold operations until further notification.'

Stornoway Town Council, Lewis District Committee and other powerful local groups had now met on the 13th to discuss the 'Iolaire Disaster Fund'. They there drew up formal resolutions and deftly lobbied both the Prime Minister and the Lord Advocate. Pressure for a Fatal Accident Inquiry boomed from the *Stornoway Gazette*, the national press and even Lewis pulpits. The political excitement was now intolerable and by 17 January, as an Admiralty Minute Sheet in File 693 makes plain, the Sea Lords had surrendered.

The Lord Advocate having been urged to hold a public inquiry was desirous, if the Admiralty saw no objection, of acceding to the requests which came in from a number of public bodies in the Lewis. He thought the best course was to hold a Fatal Accidents [sic] Enquiry which will be conducted by the Sherriff [sic]. The Procurator Fiscal will examine the witnesses and other interested parties may appear. The Lord Advocate suggested as a possibility that the shore lights might not have been in order. The proposed enquiry will be able to investigate the facts as to this. With the approval of the 2nd Sea Lord, I informed the Lord Advocate that the Admiralty would offer no objection to such an enquiry as he proposed . . .

recorded T.W.H. Inskip on 17 January. There are reluctant, initialed notes of concurrence, HIH writing on the 19th that 'in view of the rumours being circulated, I consider it very desirable that a public enquiry should be held.'

On 24 January, the Sheriff of 'Ross and Cromarty and Sutherland' had ordained a Public Inquiry under relevant Acts of Parliament of 1895 and 1906, 'within the Sheriff Court House, Stornoway, on Monday the tenth day of February 1919 at eleven o'clock in the forenoon, in regard to the wreck of HM Yacht *Iolaire* and deaths by drowning resulting therefrom.'

By month's end – if Boyle protested or objected, it is nowhere recorded in File 693 – both the Naval Secretary and the First Lord of the Admiralty had agreed not only that there be Admiralty legal representation at the Fatal Accident Inquiry, but that all their reports and minutes be sent to the Crown Office and made available to the Inquiry (though not, prudently, made public). By 4 February, the Admiralty had instructed counsel – James C. Pitman – and a Stornoway solicitor, William Ross. By 7 February, from Handsworth near Sheffield and on black-bordered stationery, even Mrs Mason (obviously concerned for her late husband's reputation) pressed for details.

As if anticipating the inevitable, Rear-Admiral Boyle had, by month's end, won, in writing, the endorsement of an expert witness, who had received and scrutinised all the secret reports, testimony and findings of the Court of Inquiry.

Minute Sheet – Admiralty

From Director of Navigation

The circumstances leading up to this accident seem to have been as follows, but the evidence is very indefinite and in parts contradictory.

The *Iolaire* left Kyle and at her normal speed of 10 knots should have cleared the North Point of South Rona (see chart cutting A) at about 9.55 pm: course was then shaped North 2 degrees East and at 12.30, when South East of Milaid Head, altered to North which was maintained until very shortly before stranding, when, according to some of the evidence, it was altered to port by entering the harbour.

The night was dark but clear with a fresh southerly wind at first, but after about 12.30 there appear to have been squalls with drizzling rain, though not sufficient to prevent Arnish Light being seen at nearly its normal distance. The course steered from 12.30 am, North, was probably for the centre of the entrance but, in the 12 mile run from that time the vessel appears (if an alteration of course was made just before stranding, as stated in some of the evidence) to have been set about 6 cables to the Eastward: which was not a large amount and does not point to careless navigation.

The entrance to Stornoway is marked by two lights close together on the port hand and when within about 3 miles of the entrance there is no other light visible by which a cross bearing can be obtained: thus a vessel enters this narrow harbour (the entrance is only 700 yards across) by means of these two lights, Arnish Point and Arnish Reef.

The cause of the accident seems to have been that finding himself to the Eastward of his intended position, the Commanding Officer altered course to pass close to Arnish beacon light but, owing to the angle at which he was approaching the harbour, this track led him close to Holm Point; being further from Arnish point [sic] light than he estimated (the brightness was probably affected by the drizzling rain) instead of clearing Holm Point the ship ran on to Biastan Holm reef.

Owing to the contradictory nature of the evidence as to the final alteration of course, it is impossible to say with certainty if this is what actually happened, but it appears to be the most probable explanation. Paragraphs (1), (2), (3) of the finding of the Court are concurred in.

J A Welsh

Director of Navigation

23rd January 1919

This 'Hydrographers theory might with advantage be put before the Sheriff,' T.W.H. Inskip would hope on 27 January. We shall later consider the mystery of the *Iolaire*'s extraordinary course. But interviewed for Gaelic radio in the 1970s about the disaster, Captain John Smith – a senior MacBrayne skipper who long served on the run between Stornoway and Kyle, in all weather conditions – did not hide his incredulity at talk of a course 'set about 6 cables to the Eastward: which was not a large amount and does not point to careless navigation.'

'I think it is too high a margin of error for safety,' Smith observed with deadly Highland understatement, 'when approaching a harbour entrance that is only 3.9 cables across.'

The Fatal Accident Inquiry opened in Stornoway Sheriff Court on Monday, 10 February 1919. So intense was the public interest there was not a spare seat to be had. Sheriff Principal MacKintosh presided; J.G. Fenton, advocate, and C.G. MacKenzie, the local Procurator Fiscal, conducted the case for the Crown. J.C. Pitman, another advocate, and William Ross the Stornoway solicitor represented the Admiralty. And a local solicitor, J.N. Anderson, founder of the enduring firm Anderson MacArthur, appeared on behalf of some of the grieving families. A jury of seven local men was empanelled: Malcolm MacLean, Point Street; A.R. Murray, Cromwell Street; George Morrison, Cromwell Street; Malcolm Ross, Francis Street; John Ross, Bayhead; Kenneth MacKenzie, Keith Street; and Angus MacLeod, Keith Street. It would be 1920 before Parliament allowed women to serve on a Scottish jury.

Mr Pitman spoke of the late Admiralty Court of Inquiry and announced its papers were at the disposal of the Crown. He was there that day only to help in every way he could, he declared, towards the fullest investigation of the tragedy. Mr Fenton responded in gracious terms, as did Anderson.

Sheriff MacKintosh said his first public act must be to express regret at the deplorable loss of over 200 men who had given brave service to King and country. He was sure all present sympathised with those throughout the island and elsewhere left desolate in these circumstances.

Lieutenant-Commander Walsh was called to the stand. He stated his office and position, and testified to the chaotic events of that New Year's Eve and the great confusion as to trains and numbers of libertymen. He related his conversation with Commander Mason. He had stated the *Iolaire* lifeboats could accommodate 100 men, though Walsh now said he had thought this optimistic. Such provision on Naval vessels was calculated for the strength of the crew, not for extra passengers. Mason had also mentioned there were 'about eighty lifebelts', but did not know the exact number.

No: no lifebelts could have been procured at Kyle. Yes: Walsh was aware of the rumours circulating, but he – himself a long-standing teetotaller – had spoken with Mason only minutes before the *Iolaire* sailed and he was perfectly sober. Walsh had not seen the vessel berth, but knew of her mishap with the pier. There had been no accident on departure. The one on arrival was a 'bit of blundering' as she had come alongside; the tides at Kyle were 'peculiar' and called for good local knowledge. He had heard it said the *Iolaire* was not a good turning ship.

Angus Nicolson, Stornoway, RNR rating and passenger, gave evidence. On boarding he had seen an officer whom he took to be Commander Mason, who invited him and others into the chartroom. He saw 'no appearance of drink' on the officer. The wind was right astern and rising, but it was not a bad night. Nicolson had gone on deck several times, the last when they were about nine miles from Arnish Light. Later some of his mates had declared, 'Well, boys, she's near in.' He had looked at his watch – it was 1.50 a.m. – and almost immediately afterwards the *Iolaire* had struck. Just before this she had altered course. He had felt change in her motion, seas coming on her beam rather than her stern. At first, he thought they had grounded by Arnish, but as rockets went up he saw the beacon on the Beasts, only 'seven or eight yards distant from the ship.'

No orders of any kind were given from the bridge or by any of the officers and crew. Nicolson had bellowed to the bridge to blow the whistle: there were 'three or four blasts'. Rockets went up, but they were not explosive, made no bang. There were no orders given about lifebelts or lifeboats. He had remembered seeing a cork life-jacket in the chartroom, and went to get it, but it was gone. About ten or fifteen minutes after grounding the *Iolaire* had slipped off the rock of impact, and swung round stern-first to the shore. He saw one of the crew at the engine-room door and told him to get her astern – she was now afloat – only to be told there was 'engine trouble.' One of the Lewis reservists had managed ashore with a heaving line, and later a hawser was taken out by it. Nicolson had made this fast to a stanchion of the depth-charge chute, and he and others had 'got ashore' by that rope. 'All that tried to get ashore did not succeed, for quite a number were washed off and drowned.' The *Iolaire* had finally disappeared about forty-five minutes to an hour after striking. He could not tell who was on the bridge, though he had heard voices and thought he saw a man signalling with a Morse lamp. He saw no flares burnt on board and the ship's guns were not fired to draw attention. In her position and with such heavy breakers, no assistance was possible from the sea, 'but if the rocket apparatus had been taken to the scene he believed many lives would have been saved.'

Kenneth MacLeod, RNR rating, Swordale, took the oath. He had seen the lights of what he took to be a 'motor fishing boat' making for Stornoway, and gave evidence to the *Iolaire*'s movements in relation to her. Both vessels were on the same course, the smaller craft ahead. The *Iolaire* at first had this boat's light on her starboard bow, but 'latterly it crossed over to port'. Soon afterwards the yacht had struck. The other boat got into Stornoway all right. Arnish Light and Arnish Beacon Light were visible for a long time before the grounding. Just a minute or so before the crash 'he and those on deck with him observed the white breakers ahead, and then became alive to their danger.' They had exclaimed, 'He's running ashore!' There had been no time to warn the bridge. After the ship struck, white, red and green rockets went up. No orders were given from the bridge. MacLeod had been washed off the rope, but a subsequent wave brought him up onto the beach.

John Montgomery of Ranish gave evidence. He had spent most of the passage in the galley, but had seen the lights of Tiumpan, Kebbock and Arnish. He had seen seas breaking on the shore and the first land he saw was on the east side of Holm Bay. A chum had observed 'he had never seen a ship running so close to the land.' The ship had altered course five or ten minutes before she struck and the wind then blew on the port beam. He had only heard one order – not to cast the boats adrift. There had been a 'slight panic', but he had seen no one jump overboard. He confirmed the rockets and the siren. He had seen one officer early in the voyage. There had been no sign of intoxication. No aid could have come from the sea. The life-saving apparatus would have been of great service. They had given no warning to the bridge on seeing the breakers because they did not think she was in danger. To a juror's question, Montgomery replied that had the *Iolaire* not altered course, 'she would have struck on the south-west point of Holm Island.'

James MacDonald of the *Spider* was called. They had just hauled their nets off the Shiants and were making for Stornoway when they saw the lights of what proved to be the *Iolaire*. He spoke of the 'changing positions of the lights in relation to each other, and said he knew that the *Iolaire* would certainly run ashore unless she altered course.' By the time the *Spider* had made the harbour entrance, the yacht was ashore and sending up rockets. They could have offered no assistance in such conditions. He had reported the matter to Admiralty authorities immediately on landing.

James MacLean, deckhand, attested he had been on the *Iolaire* since 8 February 1916. They had come to Stornoway in October. He had seen Commander Mason and Lieutenant Cotter before leaving Kyle on the Tuesday evening. 'Both were quite sober, and so was every member of the crew.' Called before midnight, MacLean took the wheel for an hour. Commander Mason had then been on the bridge but, when James MacLean was relieved by Leggett, killed a few minutes on deck and then went to the bridge for look-out duty, Lieutenant Cotter had relieved the Commander. 'Cotter was then quite sober.' Arnish light had been visible. About twenty-five minutes before she struck, Cotter had ordered MacLean below to call the watch as he expected to be in Stornoway in twenty minutes. In MacLean's absence there was no special look-out, he admitted, but Lieutenant Cotter was on the bridge and it 'was not usual to have a look-out after the harbour lights were visible.' He was below a second time calling the men and as he reached deck again, the *Iolaire* hit. MacLean did not think he had ever entered Stornoway at night since the ship was posted there. He had not seen Cotter again. 'A good while after she struck he heard Commander Mason's voice on the bridge. He was saying something about boats.' MacLean had heard no orders from the bridge. The searchlight was not used; he did not know why. None of the three guns on board was fired. All he had heard or seen was the firing of rockets and the blowing of the steam-siren.

What was his post in an emergency? MacLean was a boatman. His boat was a dinghy on the port quarter. He did not go there. He had heard no orders given to attend boat stations. Had any of the crew gone to their boat station posts? Not that he knew of. He had seen a boat in the water, tied by the full length of her painter from the side of the *Iolaire*. He had suggested to a young Lewis reservist that he should try and scramble by the painter into the boat. The boy had insisted MacLean go first. The lifeboat was in fact aground close to the shore and both of them had escaped by this painter. As far as MacLean knew no others had got off by this route. They were all trying to get off aft by the hawser.

Successive witnesses were examined – John MacKay, of Shader in Point; Alexander MacIver, Stornoway; John MacInnes, North Tolsta; Alexander Smith, Leurbost; John MacKenzie, Portvoller; Archibald Ross, Leurbost; and Norman MacIver of Arnol. All confirmed the winds blowing, the lights seen, their sudden realisation they were seriously off course; and that they had hit the rocks within a minute of seeing the breaking waves. Visibility had been good, though the night was so dark; and all the lights were clear. Any who had seen either deck officer as they had come aboard at Kyle agreed they were entirely sober.

And witnesses were emphatic: there were no orders from the bridge and neither officers nor crew had made any attempt to take the situation in hand. Only one order had been heard – 'Keep cool!' – and it had been roared in Gaelic, by a Lewisman. All the boats had been launched by passengers and on their own initiative. One had heard a call from the boat deck to keep the lifeboats alongside, but had no idea whether this had been an officer or a passenger. One Lewisman had shouted, 'Blow the whistle,' and still another had jumped to the monkey-deck and started pulling the cord. One had asked an officer to get the searchlight working, only to be told 'the dynamo was broken'. Another, like Angus Nicolson, thought he had seen an officer signalling by Morse lamp.

The Fatal Accident Inquiry adjourned for the night.

On Tuesday, 11 February, Lieutenant Hicks of Kyle was the first witness. He swore to the sobriety of the *Iolaire* officers. Two porters at Kyle station, John Beaton and Donald Campbell, who had seen the ship arrive and talked with her officers and men, saw absolutely no sign of intoxication among them.

An important new witness was Captain Donald Cameron of the *Sheila*. He had seen the little accident on the steam-yacht's arrival, but that could easily happen to a 'thoroughly careful and competent officer who was not well acquainted with the tides at that particular spot.' He had watched the ship leave 'in good order and in a thorough seamanlike fashion' and could swear to 'the perfect sobriety of the officers and men whom he saw.' Particulars of the route followed by the *Iolaire*, as reconstructed by the Court of Inquiry and the Admiralty's Director of Navigation, were laid before Captain Cameron. He judged it quite a proper one. Less than five minutes' delay in turning the ship 'could quite easily cause her to land on the rocks where she did.' He and the *Sheila* had left Kyle 'about an hour' after the *Iolaire* and never saw her lights until after she had grounded. He had reported the wreck immediately on making Stornoway. He had only asked men aboard the *Sheila* to make up her complement that night. 'I saw absolutely nothing wrong with any person or thing about the *Iolaire*.'

Donald MacDonald of Upper Bayble and John MacPhail of Doune, Carloway, spoke of the ship's course, which they had thought quite correct until shortly before grounding. Neither had heard any orders from the bridge.

Various members of the Coastguard gave evidence on the turgid saga of the life-saving apparatus. John MacSween of Stornoway had been roused at three in the New Year morning. Hugh Munro and Hugh MacLeod corroborated his account of the call-out and the wretchedly slow progress to Holm. William Barnes stated he had got a message from the Battery about three in the morning. The Chief Divisional Officer, Boxall, had been summoned and arrived at 3.40. Boxall confirmed this. He had only been in Stornoway for a month before the accident and had not met the crew or had any exercise with them. He thought the LSA had left the barracks not later than 3.50 a.m. Two survivors had reached the Battery before he left 'and had told him the apparatus would be of no use as by that time the ship had totally disappeared.'

Lieutenant-Commander Morris, RN, gave the bald figures – 284 on board and 79 survivors; 138 bodies recovered to date, of which 130 had been identified. There were still 67 missing. In the Royal Navy a ship carried only sufficient lifebelts for her complement. The *Iolaire* had wireless 'but the operator had found it impossible to get it to work on the night of the accident.' Dr Owen, RN, confirmed he had examined all recovered bodies. John MacLean and Duncan MacKenzie of the Lifeboat Institution confirmed they had been summoned that morning, but Rear-Admiral Boyle had told them the lifeboat could be of no assistance in the wreck's position. Commander Bradley, who had sailed with Cotter, delivered a veritable eulogy, insisting he was 'one of the last men to have no look-out' and 'a most careful and cautious navigator'.

William Saunders, signalman at the Battery, had been on duty from 12 to 4 a.m. on 1 January. He had seen the lights of two steamers approaching Stornoway. One entered harbour, one seemed to go 'further east than she should have done.' She showed a blue light. He had reported this as 'a ship wanting a pilot'. Then she showed another; and later a red rocket. Saunders had then tried to communicate with her by Morse lamp but found this impossible: the headland was between his sightline and her bridge.

Robert Ainsdale confirmed Saunders had reported 'a vessel off Holm Head, showing a blue light, wanting a pilot' at 1.50 a.m. He had telephoned the Rear-Admiral and was told a pilot would be sent out. At 2.15 he saw a red rocket. Again he had telephoned Boyle, who said that all assistance would be sent out and also ordered the despatch of the LSA. Ainsdale then sent for Barnes at the Coast Guard, who came promptly. He thought the LSA had left the barracks by three o'clock. Lieutenant Wenlock gave evidence. He had been about to sail out as pilot-boat on Boyle's instructions when he saw a green rocket go up 'from a ship somewhere outside' and had run to report this to the Rear-Admiral. He got no answer immediately and did not wait.

Proceeding to sea, Wenlock had seen more rockets, but when the *Budding Rose* reached the scene they could see no signs of the ship though they were 'practically alongside of her when the last rocket went up.' He returned and had reported her exact position shortly after three. Help from the sea would have been quite impossible.

No further evidence was led. There are some striking omissions: Leonard Welch, for instance, the wireless operator – who saw more of the officers than anyone else – was not called to testify. (His evidence on the state of his equipment would have been a public sensation – wireless, its adequacies and its 24-hour manning, had been a minutely covered aspect of the *Titanic* disaster.)

Mr Fenton addressed the jury. In such a disaster speculation was inevitable and there had been 'rumour of the most cruel and base kind.' But every witness in the case confirmed the sobriety of the officers and – representing the public interest – he was quite satisfied none of the crew was in the least affected by liquor. Up to within a few miles of Arnish, he had no criticism of the navigation of the *Iolaire* – Captain Cameron had stated in evidence her course was neither out of the ordinary nor improper – but, when the Navigating Officer had seen another vessel obviously making for Stornoway, and on a different course, he should have slackened speed. He should also have had a competent look-out. Witnesses disagreed as to whether the *Iolaire* had altered course just before grounding, but a few minutes earlier would have made all the difference between safety and shipwreck. All had agreed there were no orders issued after she struck; no attempt made to control the men, nor to get the crew to their boat-stations. The arrangements for using the life-saving apparatus needed 'drastic revision'.

Mr Pitman insisted the Admiralty 'welcomed the Inquiry' and he only wanted to help Mr Fenton as much as he could to make it as full and complete as possible. He had thought it fair to Lieutenant Cotter's memory to call Commander Bradley, who had attested to his care and competence. There was no insinuation against Commander Mason –

'there could be none against such a man.' The Admiralty, concluding their Court of Inquiry, had not thought fit to find fault of any kind. On this evidence, the jury could not be justified in apportioning blame either. 'An over-running of the course for a few minutes, a mere error of judgement, was sufficient to account for what had happened, and to say on account of that error of judgement the Navigating Officer was incompetent was unfair to his memory.'

Mr J.N. Anderson, on behalf of survivors and some bereaved families, was pleased aspersions on the dead officers had been dispelled by the evidence led. Some press reports had asserted confusion had arisen on board because the RNR libertymen did not understand orders given in English. No one who had heard their testimonies could say they were unable to understand English. 'As a matter of fact,' said Mr Anderson, 'no orders were given.' There had been gross carelessness and mismanagement on board after the vessel struck, besides 'mismanagement' of the life-saving apparatus.

Sheriff-Principal MacKintosh summed up. It was unfortunate that the accident had happened at a season so associated with conviviality. He was glad some of the lying rumours about the officers in this connection had been dispelled. It had been proved there was no truth in those rumours, and he felt it proper to remove such a reflection from their memory and from the minds of their relatives.

The jury must be chary about assuming the position of censors and distributing blame. A great many casualties at sea must forever remain enigmas. Those who could speak to the cause of the wreck had perished with the vessel. Sheriff MacKintosh attached great importance to Captain Cameron's testimony. From that evidence it seemed probable there was not much wrong with the course of the *Iolaire* and keeping that course a few minutes too long near the entrance of Stornoway Harbour had caused the disaster.

The evidence of a good many witnesses certainly gave the impression that, after grounding, 'there was something lacking in the matter of order and supervision and control', but the jury should bear in mind evidence from two witnesses that both Lieutenant Cotter and Commander Mason did give at least one order from the bridge relating to something about boats. All agreed no effective help could have been given from the sea, but there was doubt as to whether the life-saving apparatus arrangements on shore were satisfactory.

The jury listened, retired, and took just over an hour to return.

The jury unanimously find (1) that about 1.55 am in the morning of 1st January 1919, HM Yacht *Iolaire* went ashore and was wrecked on the rocks inside the 'Beasts' situated near Holm Point in the Parish of Stornoway, Island of Lewis and County of Ross and Cromarty (2) that the deaths resulting therefrom amounted to 205 and that the cause of death was suffocation due to submersion (3) that the Officer in Charge on said occasion did not exercise sufficient prudence in approaching the harbour (4) that the boat did not slow down, and that a look-out was not on duty at the time of the accident (5) that the number of lifebelts, boats and rafts was insufficient for the number of people carried and that no orders were given by the Officers with a view to saving life and further that there was a loss of valuable time between the signals of distress, and the arrival of the life-saving apparatus in the vicinity of the wreck.

The jury recommend (1) that drastic improvements should be made immediately for conveying the life-saving apparatus in future to give assistance in cases of ships in distress (2) that the Lighthouse Commissioners take into consideration the question of putting up a light on the Holm side of the harbour (3) that the Government in future will provide adequate and safe travelling facilities for Naval ratings and soldiers.

The jury desire to add that they are satisfied that no one on board was under the influence of intoxicating liquor and also that there was no panic on board after the vessel struck.

Rider to Verdict

The jury recommend to the Carnegie Trust and the Royal Humane Society, Seaman J.F. MacLeod for some token of appreciation of his conduct in swimming ashore with the line by which the hawser was brought ashore by which means many lives were saved.

The jury also desire to extend their sincerest sympathy to those who have lost their relatives in this regrettable disaster and also express their appreciation of the hospitality shown to the survivors by Mr and Mrs Anderson Young, Stoneyfield Farm.

Chapter Seven

'My heart is so full of him –
and my loss of him so great'

Agus thuirt ise riu, Na abraibh Naòmi rium: abraibh Màra rium; oir bhuin an t-Uile-chumhachdach gu ro-gheur rium. Chaidh mi a-mach làn, agus thug an Tighearna air m' ais mi falamh. Carson, ma-ta, a their sibh Naòmi rium, agus gun tug an Tighearna fianais am aghaidh, agus gun do chuir an t-Uile-chumhachdach fo àmhghar mi?

Rut i: 20–21

And she said unto them, Call me not Naòmi, call me Mara: for the Almighty hath dealt very bitterly with me. I went out full, and the LORD hath brought me home again empty: why *then* call ye me Naomi, seeing the LORD hath testified against me, and the Almighty hath afflicted me?

Ruth i: 20–21

Such respect as one might retain for Robert Francis Boyle, Rear-Admiral, Commander-in-Chief at Stornoway, third son of the Earl of Shannon, does not survive his bitter response to the Fatal Accident Inquiry – as unveiled in pitiless detail, fifty years later, in File 693.

J.C. Pitman, the silky-tongued advocate, was no less a sore loser. 'It may interest you to know that I am satisfied it was just as well that the Admiralty were represented,' he mewed in a 13 February note to Thomas Carmichael, an Edinburgh official of the Royal Navy. 'From the talk in the place and questions put by individual jurymen, it became clear to me that the latter would be only too ready to give a verdict which would reflect seriously on the Admiralty as responsible for the officers and crew, the taking of so many libertymen on board with half a crew, and no lifebelts, etc., and being in some way responsible for providing men for the life-saving apparatus and the lifeboat . . . as there were enough fishermen in the harbour to man the lifeboat two times over, the jury seems to have thought it better to say nothing about it. The whole Inquiry seemed almost like a criminal trial with the Admiralty in the dock – and the difficulty that any questions put by me were looked upon as an attempt to shift blame from off official shoulders.'

'I hope,' Pitman concluded, 'that the jury's verdict will now be accepted by the population as being the worst that can be said in regard to the responsibility of the Naval Authorities, but I doubt whether it will.'

Boyle, too, seethed. 'The jury apparently had little technical knowledge on nautical matters and appeared to be prejudiced against the Royal Navy generally. No Nautical Assessor was called for the Crown. The jury must have seen by the evidence that the promptest possible action was taken by the Naval authorities, though no mention is made of this in the report which appeared in the press.

'The jury seems to have been ignorant of the life-saving apparatus equipment – that the cart is not easily handled on bad roads and under unfavourable weather conditions – that the life-saving apparatus is under the Board of Trade – and that it is only a 'Whip' equipment. The cart was taken by naval ratings from the Depot the whole way, and the horse did not appear until the return journey. The Court in addressing the jury seem to have had an idea that the life-saving apparatus can be called out on the lines of a fire engine working from a first-class fire station.'

And he was hurt 'no witnesses were called by the Sheriff to show the steps which were taken by the Naval Authorities to render assistance'. The Rear-Admiral had even formulated a conspiracy-theory about Surgeon-Lieutenant Thomas Owen – 'no questions were put to him as to how and at what time he reached the scene of the wreck; it would appear that such questions were pointedly omitted as they would have brought to light the difficulties in obtaining transport for stretchers, blankets, medical comforts, etc.'

The Rear-Admiral wanted vindication. 'I submit that if the Minutes of the Enquiry [sic] are available, and after perusal of the evidence therein, Their Lordships may think fit to make a statement in the public press concerning the prompt action taken by the Naval authorities, all orders for which were personally issued by me, to dispel a feeling which the public appear to entertain that the Naval authorities were inactive.'

His London superiors studied the Fatal Accident Inquiry, and these pompous epistles, with an anxiety all the greater as Dr Murray, Western Isles MP, had now tabled a Parliamentary question, asking if the First Lord's attention had been drawn to the findings on the loss of the *Iolaire*, 'in which a strong recommendation had been made that proper provision should be made for the safe transit of Naval ratings on leave to the Western Isles.' The Admiralty quickly sensed the elephant-trap: any assurances of better arrangements in the future would only emphasise the disarray on 31 December and be practically an admission of culpability. Dr Murray's question ping-ponged from desk to alarmed desk.

Some bucks could readily be passed. 'The question of the life-saving apparatus on shore is for the Board of Trade, and that of the Harbour Light proposed for the Northern Lighthouses Board. These questions will presumably be sent to the proper departments by the Scottish Office,' resolves an Admiralty Minute of 31 March 1919.

'Submitted whether a statement such as is desired by RA Stornoway should be attempted. It seems not improbable that a further question will be put in the House of Commons and the evidence at this Public Enquiry [sic] was obtained in view of this contingency. Meanwhile it may not be considered advisable that the Admiralty take the initiative in re-opening public controversy in order to make clear what may be regarded as the minor incidents of the disaster . . . Rear-Admiral, Stornoway, to be informed that it is not considered desirable to re-open this matter and that no statement will, therefore, be published in the press.'

An icy note on 4 April 1919 – BY COMMAND OF THEIR LORDSHIPS – broke the news to Rear-Admiral Boyle.

And, of a piece, the contents of File 693 reflect the moral disengagement of men to whom Highlanders were cattle.

The Royal Naval Reserve depot at Stornoway would close later in 1919. The site at the Battery is now dominated by the local power-station. Rear-Admiral Boyle died in 1922, still only fifty-nine years old.

In Tolsta, a protracted and dangerous diphtheria stalked the village for much of 1919 – linked, at last, to the dubious school latrines. But that summer, wretched as villagers felt and heavy as were many hearts, there was determined effort to cheer, at least, the young.

Lewis Peace Celebrations – North Tolsta

On Saturday, 19th July, there was a peace celebration picnic for the school children and others, held at Garry. As there was a funeral in the village that day, the festivities did not begin until 2 pm. Over 200 children marched from the school to Garry, about 2 miles distant. Before the sports started, the senior pupils amused themselves by gathering water lilies on Loch na Cartach. The sports commenced about 5 pm and many of the children took part in the following events – 300 yards, sack, three-legged, wheelbarrow, donkey and marathon races, egg and spoon races, tug of war, etc. There were over 100 prizes given and every child present received a coin in memory of the occasion. Mr Dunlop gave out the prizes. At the close, Mr MacDonald, Schoolmaster, made a short speech suitable for the occasion. After the votes of thanks, the children were formed into procession and marched homeward, highly delighted with the day's outing . . .

Tolsta had sent 231 men to war: fifty-two had died, eleven of them with the *Iolaire*.

Below a report in *The Scotsman* of the first day of the Fatal Accident Inquiry proceedings appears a little news item.

For Relief of Sufferers

Towards the Iolaire Disaster Fund, Stornoway, Mrs Kennedy-Fraser, Edinburgh, yesterday dispatched a cheque for £100, this being the clear profit of a recital of the 'Songs of the Hebrides' given by her in the St Andrew's Hall, Glasgow, on January 21, with the help of her daughter, Miss Patuffa Kennedy-Fraser, and her sister, Miss Margaret Kennedy, and of Miss Ruth Waddell, cellist.

There is an inveterate public instinct, today as much as then, to react to tidings of dreadful human calamity by reaching for our wallets. The need to do something – anything – is felt, and the easiest response, at once the laziest and the most gratifying, is to donate money, as if hard cash might somehow neutralise catastrophe and assuage the pain of an empty chair, a desolate bed, a shattered life. Then, as now, a great pile of cash can cause as much trouble as relief.

In joint session on 13 January 1919, the Lewis District Committee, Stornoway Town Council and the War Pensions Sub-Committee for the Lewis District had agreed to mount a 'Public Appeal' for 'funds to supplement official allowances and grants payable to Widows and dependents of the men lost in the *Iolaire* disaster . . . Thereafter the following gentlemen were appointed a Committee to draw up a Public Appeal and to make the necessary arrangements in connection therewith viz. The Chairman, Provost MacLean, Bailie MacLeod, Ex Provost MacKenzie, Dr Murray MP, Mr Angus MacLeod, Rev. George MacLeod, Rev. R. MacLeod, Mr A.R. Murray, Mr J. McR [MacRitchie] Morrison, Mr N. Stewart and Mr John MacKenzie, School Board Clerk, Tarbert, Harris, with powers to add to their number. Provost MacLean was appointed Chairman of the Committee and Mr Angus MacLeod, Secretary.' That business concluded, the great and good of Lewis had then passed a still more robust resolution calling for a 'Public Inquiry by an Independent Tribunal' into the loss of the *Iolaire*.

The failure to appoint a single woman to this committee is noteworthy. But the immediate success of the appeal is undoubted. Donations were already amassing – £1,000 from Lord Leverhulme; £200 from the Highlands Society, Edinburgh; £500 forwarded from the Grand Fleet Fund, 'and other remittances'. Weekly the *Stornoway Gazette* continued to print acknowledgement of monies great and small: from all over the British Isles, from dominions such as Canada, from assorted public eminences, and from the many self-help organisations for Highland people in the cities of the south.

There was, for instance, a great gathering in the King's Theatre, Glasgow, by the united Highland societies of the city – the Lewis and Harris Association, the Skye Association *et al.* – where, before the curtain arose on an ensemble of choirs and soloists, pianists and fiddlers, playlets and *tableaux vivants*, the 'celebrated Shakespearean actor' Sir John Martin Harvey recited a 'Prologue Written by the Players' – the work of Neil Munro, a Glasgow-based journalist from Argyll now mostly remembered for his jolly Para Handy titles, originally published under the pseudonym 'Hugh Foulis', but who in 1919 thought himself a serious novelist.

April has come to the Isles again bleak as a lover;
Shaking out bird-song and sunshine and soothing the tides;
April has come to the Hebrides, filled them with frolic,
Only in Lewis of sorrow, bleak winter abides.

Always they went to the battles, the people of Lewis,
And always they fell, in the wars of a thousand years;
Peace never to Lewis brought Springtime of joy or of season –
The wars might be won, but her women were destined to tears!

That is, today, why in Lewis the lark sings unheeded,
The sparkle of waves in the sea-creeks gladdens no eye,
No dance to the pipe in the croft, and no mirth in the sheiling,
Cheerless and leaden the hours of the spring go by . . .

The gathering nevertheless raised a considerable sum.

Meanwhile, calamity continued to unfold in Lewis; a virus that would slay far more people than the Beasts of Holm. We have already mentioned the 'Spanish 'flu'. The name is a misnomer, won because the Spanish press – in a country which stayed neutral throughout the Great War – had no media censorship and gave the outbreak sensational coverage when the influenza arrived in November 1918. It had, in truth, first appeared in Fort Riley, Kansas, in the US in March and many assert it was brought to Europe by the thousands of American troops. And it has been estimated – though it is unlikely ever to be proven – that in the months this global influenza raged, from March 1918 to June 1920, it killed between 20 and 100 million people, even the lower figure equating to a third of the entire population of Europe and more than twice all the fatalities of the First World War.

It was a peculiarly severe 'Influenza A virus strain of sub-type H1N1', and in most communities it was caught by 50 per cent of the population. It was marked by extraordinary violence of symptoms and, because it attacked the tissues by 'cytokine storms' – gross overreaction of the immune system – it was especially dangerous to young adults, who have the most vigorous immune systems of all. Most finally died of the secondary complication, bacterial pneumonia, but the virus in many cases caused fatal haemorrhage from the mucous membranes in the nose, stomach and guts; or deadly oedema in the lung.

Even the humdrum 'seasonal 'flu' can still kill; this 1918 strain was exceptionally lethal. Between 2 and 20 per cent of all sufferers died: 99 per cent of the casualties were under 65, and more than half were young adults between the ages of twenty and forty. Famous victims included the English composer Sir Hubert Parry, German economist Max Weber, the first prime minister of South Africa and the president-elect of Brazil; notable survivors (who caught the 'flu but recovered) count David Lloyd George, Kaiser Wilhelm himself, the young Walt Disney, the actress Mary Pickford, the serving President of the United States, Woodrow Wilson, and a future one, Franklin D. Roosevelt.

A record of Ness burials shows starkly the scale of the pandemic in 1919. By late March the 'flu was rampant – on the 22nd, and on just that one day, five people were buried at Swainbost cemetery – and thereafter people died daily, 1 April alone seeing another five interments. In 1919, Ness buried nineteen children of five or under, and twenty-four additional people of 25 or under. But the scale of the outbreak is most evident in flat, overall figures. In 1918, there were thirty-nine burials in this community of Ness. In 1920 – the 'flu had, by then, burned out – there were forty-six. In 1919, *excluding* the *Iolaire* fatalities, there were ninety-five interments at Swainbost Cemetery; when those men are added, 118 people died in this one corner of the island in just one year, and nowhere in Lewis and Harris escaped comparable desolation.

Some comment on the strange manner in which Lewis people seemed to 'park' the *Iolaire* disaster for much of the twentieth century; or suggest that, in some collective form of post-traumatic stress disorder, the Isle of Lewis has refused then and since to come to terms with what happened on the Beasts of Holm. No one should make such fatuous pronouncements without pondering, long and hard, the biggest killer of 1919.

The well-meaning directors of the Public Appeal could not have anticipated the astonishing rout of Lord Leverhulme, general political mismanagement of the national post-war economy, the circumstances that would bring both mass-emigration and dreadful destitution to Lewis and Harris in 1923 or the wider hardship that would finally negate their fund-raising. But they might perhaps, amidst the ongoing slaughter of the Spanish 'flu, have anticipated what we would now dub 'compassion fatigue'; they should certainly have expected the natural reaction of many, many Great War widows in the Long Island who had been bereft by shell, bullet and torpedo rather than woeful seamanship – and would get no special cash award – and the very broad definition of 'dependent', especially in a pre-welfare state world and in a community of staunch extended family. Today, only a widow and her children would expect to benefit from such charitable endeavour: in 1919, self-defined *Iolaire* dependents included siblings, parents and even grandparents – all living in the given casualty's household and, in many instances, beyond it. Instinctive, guilty public response had raised a great sum – over £9,000 by late 1919 – only to trap its directors in a draining moral maze.

In a large foam-lined cardboard box, kept in the controlled conditions of a warehouse on the outskirts of Stornoway, is a poignant collection of documents. Between the marbled boards of a busted ledger – the front cover detached years ago, and has since served as a makeshift paperweight – sits, of course, the accounts of monies disbursed, donations received. But atop that are piled copies of a single foolscap form, sheet upon sheet upon sheet, filled out by dozens of mourning hands. Many posted a letter besides, often on the black-bordered mourning paper still then fashionable, now anchored to each form by a rusting pin.

These are the papers of the Iolaire Disaster Fund, which no author has previously seen. And, even at ninety years' remove, much of the intimate and at times heart-rending detail has to be quoted sensitively: at least two infants whose names appear on these foxed files of 1919 are still living.

More than anything else, the simple act of leafing through these bits of paper – sheet upon sheet, for drowned upon drowned – at last brings home the sheer scale of this tragedy in a manner that no clutch of little headstones or a bald list of printed names can ever do. I am not long at the chore before, to my blinking and embarrassment, I have wet eyes;

though that hardens shortly to anger as the great sheaf of elegant printed forms gives way briefly to a sequence of ugly, mulberry, mimeographed typescript – the papers belatedly issued to families in Harris. Neat printed things had gone immediately to the grieving in Lewis – all her *Iolaire* losses were published, with addresses, in the *Stornoway Gazette* on 10 January 1919; and to relatives of all lost crewmen of the yacht.

' "Iolaire" Disaster', each form is boldly titled, and below is the mannered invitation, 'The Committee of the "Iolaire" Disaster Find will be pleased if you fill in the information asked for in this Form and return the Form, as soon as possible, in this enclosed addressed envelope.'

A succession of dotted lines seek the name of the 'Late' such-and-such, drowned on the *Iolaire*; the 'Name of Dependent', 'Address', 'Relationship to Sailor', 'Official Number of Deceased', his 'Rating' and – in a table of boxes for the claimant, 'Relationship to Deceased', 'Name', 'Age', 'Occupation or School', 'If in Employment, Wages Earned' and 'Particulars of Members of Household'.

There follow more dotted lines demanding, under 'Income', 'Total Amount of Allotment and Separation Allowance received in respect of Deceased', 'Total Amount of Allotment and Seperation [sic] Allowance received in respect of any other member or members of household', 'Total of other source of Income', and a declaration – 'I certify that the above statement is true to the best of my knowledge and belief' over a last space for signature and date.

Deckhand Donald MacLeod of 3 North Tolsta, only twenty, left a family in typical straits – his seventy-four-year-old grandmother, a forty-four-year-old aunt, her four-year-old son and a brother of sixteen. In his last months of service, they had received a weekly 'allotment and separation allowance' of 12s; which continued, under Great War provision, for six months after a serviceman's death. Mrs Margaret MacLeod, grandmother, had besides her 'Government old age pension' of 5s 6d a week. She writes Angus MacLeod, secretary of the Fund, with dignified hand, hoping she has correctly given 'all particulars required.' Her anxiety is evident (and on such a budget, no wonder) as 'one of the separation allowances has since been withdrawn.'

Margaret Murray's husband, Evander, of 45b North Tolsta, had been forty-five. She is thirty-six, a 'housekeeper', and has a daughter Catherine, 19; she notes honestly in the margin that Catherine is 'but stepdaughter of the dependent'. The girl is a 'domestic servant'. Margaret Murray has no wages; Catherine has 'none at present – in delicate health.' Mr Murray's Admiralty allowance of £1 a week – married men with one or more children had more generous provision – was their only cash income, and would cease in July. 'My father Evander Murray was drowned in the *Iolaire* disaster,' writes young Margaret in her accompanying letter of 10 February, 'and I am his only child. My mother died 5 years ago and I am living with my stepmother and we both were dependent on my father's wage . . .'

A sailor in a village near Stornoway had been widowed. He left a 4-year-old daughter in the care of her guardian, his fifty-five-year-old mother-in-law. Their allowance was just 10s a week. Years afterwards, there would be a protracted correspondence over this girl, reflecting the desperation of the times, the very limited opportunities for women and her alleged frailty. 'The guardian of Miss ____ whose father was lost in the *Iolaire*,' writes the village schoolmaster in January 1933, 'was seeing me last night complaining that her granddaughter, the said Miss ____, received nothing from the Iolaire Fund for some time. The girl is about 18 years of age, and is of a weak constitution; and I would be obliged if you looked into the matter at your earliest convenience . . .'

'On 1st July last,' wrote the long-suffering secretary in response the following day, 'Mrs ____ received the usual allowance from the Iolaire Fund. Mrs ____ was not a dependent of the father of Miss ____ and merely received the allowance as guardian of the child. Miss ____ having reached her eighteenth birthday in May last the allowance according to the decision of the trustees then ceases. I shall, however, be glad to have from you any information regarding the girl's circumstances so that I may submit same to the Committee.'

'In the special circumstances of the above girl,' rallied the teacher a fortnight later, 'I would, from personal knowledge of the case, recommend the relief to be continued. The girl has always been weak, and on the advice of the late Dr Macrae she did not take up any work. She is always ailing and if necessary can supply med. evidence in support of this. Kindly lay the matter before the trustees . . .' On 18 April 1933, the Secretary has a grand letter from the local Free Church minister. The agitating old lady was evidently formidable. 'I am instructed by the grandmother of Miss _____ to state that this orphan girl is in ill-health and under medical treatment for some time and thus those who have to do with the administration of the Iolaire Fund should continue pension in her case as hitherto her grandmother gets 10/- per week only and without Croft or anything else. They can hardly exist on this small income. I hope and trust this case will be favourably considered.'

A special payment of £5 was duly sent by cheque to the young woman on 1 August. 'I was also asked to say that the payment of this special grant is not to be looked upon as forming a precedent,' advises the Secretary in his austere note of defeat.

The Committee had not envisaged pressure from all sorts of second-hand 'dependents', once an old widow died or an orphan came of age; could certainly not have foreseen the hunger for cash, any cash at all, in the hungry Twenties and Thirties, whatever stratagems were deemed essential by some islanders to secure it. The trustee's worst case fell in one community far from Stornoway in the late 1920s, when the aged father of an *Iolaire* casualty died and his provision naturally ceased. A surviving son and daughter, who had shared the family home, then made a frightful fuss. Letters flew back and forth, a solicitor nearby was contacted to engage local spies, and on information received it was decided to give a small annual grant to the woman (yet another of 'fragile health') but not to the brother; the lawyer's surveillance confirmed the chancer had his own Navy pension and was in paid employment besides.

Many of the drowned had left large families. John Morrison, 10 Coll, forty-four, was survived by a forty-one-year-old widow, Kate, and eight children. The beseeching (and most articulate) letter his Kate had to pen must have been a humiliating experience.

Sir,

I beg to apply for a grant from the Iolaire Disaster Fund. My husband, John Morrison, was drowned in the *Iolaire* Disaster. He was at the time in the Royal Naval Reserve. Before the war he was a fisherman.

My family numbers nine persons, including myself. Their ages vary from 18 years to 1 year. I myself have been under medical treatment by Dr MacKenzie for the last three years, and for the past eight months have been practically confined to bed. I am thus quite unable to do any housework. The result of this is that my two eldest daughters, aged 18 and 16 respectively, have to be always at home to look after the house & the younger children.

We have a small croft but owing to our having to pay for the ploughing and other heavy work done on it, it is a great expense. In the same way our peat-cutting costs a considerable amount, owing to there being no man at home to look after it. In fact I have had to borrow money for the working of the croft and peat-cutting this year. While my husband was in the navy we were receiving the sum of 52/6 per week but lately this has been reduced to 43/4 by the ministry of pensions. My two eldest daughters who are both over 16 are deemed to be capable of supporting themselves and are not allowed anything, altho' as I have pointed out owing to the state of my health & the claims of my young family they are both needed at home and are thus unable to go out to work for themselves.

Dr Mackenzie will, I'm sure, give information of my ill-health if such is thought necessary. Rev. R MacKenzie will also give any other information necessary in support of my statements. I shd. be obliged by your bringing my case before the Trustees as soon as possible.

Yours faithfully

Mrs John Morrison

The Committee may have wondered why two grown daughters had to be in full-time attendance at home – three of the other children were over five years of age – but an award was nevertheless granted. (The sum in this case is not indicated, but typical Disaster Fund assistance ranged between £7 to £9 10s a year.)

John MacAskill, Lighthill, Back, had been twenty-seven and single. He left an unemployed father of sixty-two, a fifty-seven-year-old mother, a twenty-nine-year-old sister, Catherine, 'in ill health for a long period', and two younger sisters, Anne and Jane, 'at home'. Their entire income was £2 6d a week. His father, Murdo, could not sign his own name, and attests with a cross. 'This family are in very vulnerable circumstances,' their minister has written below. 'The only two sons in the family are gone – one killed in France and the other was drowned. RMK. Manse.'

The body of Angus MacDonald, 24, Port of Ness, would yet be recovered. His father, Angus, was a 57-year-old crofter and an invalid; his mother, Margaret, was fifty-four. Two sisters in their twenties were unemployed 'fish workers' – *clann-nighean an sgadain*, the girls who followed the herring-fleet around the British Isles, gutting for hours on end and living in draughty lofts. The industry was already toiling; the great shoals had pulled one of their occasional disappearing acts and the Russian Revolution had closed the main market for salt herrings. Two younger brothers were still in RNR service and twelve-year-old Annie, youngest of all, was in school. The MacDonalds got £2 10s a week. 'As I have not so far received any contribution from the Emergency Fund named above,' scribbled Mrs MacDonald on 26 August 1919, 'I should feel much obliged if you will be good enough to investigate whether the matter has been overlooked. My son's name was Angus MacDonald and he served on HMD *Primrose* immediately before his last leave home. Our home circumstances are such that I cannot very well understand why relief should be withheld. Thanking you in anticipation. I am, Yours faithfully, Mrs A MacDonald.' Her husband died on 8 January 1920; she spent thirty-two long years as a widow.

The body of Angus Morrison, a twenty-year-old rating from 10 Eoropie, was never found. His ageing parents had to raise his three younger sisters and two little brothers on just 22s a week.

To A Munro
Iolaire disaster fund
21-6-21

Sir,

Sorry I have to write you for it is hard lines for me to write for this purpose. Hopping that you will look through the matter for the like of me, who lost his son aboard HMY *Iolaire*. Hopping in such hard times as this that you will help me, I think when the money is lying at hand for the purpose of helping the likes of me that I should get as much as would keep us from starving. There is know way that I can look at for help since I lost the only help I had, and if I had not right for the money lying at hand which was subscribed through the country which I cannot see the reason that it is kept back I would not write you.

Hoping you will write me as soon as possible from your faithfully John Morrison.

'Allowance as usual,' Mr Munro has inscribed on 2 December 1921.

At 6 South Bragar, Murdo MacLean, forty-two, was survived by his wife, Effie, four sons aged from twelve to one, and an eight-year-old daughter, Malina. Their allowance had been £3 2s. In November 1921, the Secretary of the Lewis School Management Committee, D. MacKay, applied to the Disaster Fund for additional help in this case: little Malina had to go away for an operation on her left hand, 'the thumb and little finger of which are tied down by contraction caused by a burn in infancy.' '£5 granted', Mr Munro has noted on this letter, '2/12/21'. Murdo's thirty-seven-year-old brother, John, 17 South Bragar, had also died on the *Iolaire*, and their mother Ann, seventy-seven, applied – and was granted – assistance in February 1920. She could mourn at John's grave; Murdo was never found.

The bittersweet tensions in one family were evident at 6 Kirivick, Carloway. John MacLeod, nineteen, is buried at Dalmore. He left a widowed father, Malcolm, fifty-three, and six brothers and sisters: the oldest, Effie, was twenty-one. None was in employment, three were still in school, and they had no cash income whatever. 'Living depending on my own labour,' Malcolm has written in beautiful script. In the attached note he told the Committee, 'I lost my wife the mother of my family some years ago and was about to marry a second wife when my son John joined the service last spring. So he deprived me of his allotment and separation allowance, not that I was independent of support though he did not grant it to me. Malcolm MacLeod.' A firstborn son especially may have a fraught relationship with his father, but we know nothing of the dynamics here and John had no chance, of course, to tell us his side of things. The straits of this household were evident: £6 was awarded.

Some cases make one shudder. An *Iolaire* casualty in one northern Lewis village left a fallen woman – a twenty-eight-year-old 'fiancée' with their 2-year-old illegitimate daughter and her sixty-six-year-old mother. Or had he? We have only the young woman's word for it, and lest the child yet live or the drowned rating was in truth innocent, they should not be identified. 'There was no allowance in this case,' one 'JMM' has scrawled. 'Rating helped child and mother otherwise.' The Disaster Fund Committee decision is not recorded. Given the impossibility of establishing paternity and the mores of the time, it is unlikely the girl received a penny.

At Earshader, Christine Smith, in a clear hand, completed a claim for herself, her 4-year-old daughter and baby son. They had enjoyed but 21s weekly from the Admiralty. £7 was awarded. 'Dear Sir,' she would write on 18 March 1920, on black-edged notepaper, 'I have not got my own birth certificate but my age is 32 years last February. Yours truly Christine Smith.'

At 16 Leurbost, Donald and Mary MacKenzie were yet another couple, well in their sixties, shattered by the loss of two sons, Alexander and John. They shared the house with two daughters in their twenties and had had an Admiralty allowance of just 12s a week. 'See back,' someone has scrawled, and there we find a later adjustment. 'Donald MacKenzie dead. There being two sons lost in the disaster, it was decided at a meeting on 2/12/21 to transfer this payment to the widow of above, in addition to the one she receives for the other son.' In South Lochs, forty-four-year-old Angus Montgomery had also left a great brood: his wife, still only thirty-five, and their eight children. They had a weekly pension of £3 14s 2d. Grant was awarded.

Murdo MacIver, forty-nine, 36 Lower Bayble, had left a forty-eight-year-old wife, a sickly father-in-law and – the form has been completed most confusingly – between four and eight children. By March 1921, Mrs MacIver – by then so crippled with rheumatism she could not even dress herself, and so tended by her daughter, 'without wages or clothing' – was forced to conduct mortifying correspondence with the Disaster Fund. 'I cannot do anything to her by way of wages after paying Doctor for medicine. She must leave me for [domestic] service, and I am left alone with the small children . . .' Ann MacIver hoped 'you will grant some wages to my daughter . . . or increase my own allowance which shall be paid to her . . . As this is an urgent case which need careful attention will you be as kindly to answer by first post . . . The situation can be testified by Rev. George MacLeod and Dr Stewart who is attending me.'

Days later, forced to state the household income all over again, 'Now after paying Doctor's fees and medicine and other prescription, the family is suffering, for they had to sell two Cattle to pay part of my expense, as for my daughter who is attending me she has run out of cloth and books and many other thing, even food, as the pension cant pay for everything.' Still more embarrassing details of her illness and dependency follow, and her desperation is evident. 'Now Sir I would rather see one of the Trustees and speak to him personal than writing, as he would see matters as they are. I am, Sir, yours obediently, Widow Murdo MacIver, 36 Lower Bayble, Point.' Thirteen shillings was duly added to her Disaster Fund pension; but the wretchedness of such entreaty can be imagined.

Petty Officer Donald Murray, 43 South Shawbost, had been forty. He had left a widow and three daughters; a fourth was born precisely a month after his death. Murdina Murray filled out the details only a week after her confinement, five weeks after her widowhood, on 8 February. With a literal understanding of the question, 'Total Amount of Allotment and Separation received in respect of Deceased,' Murdina did the fraught arithmetic and wrote, '£250 since outbreak of war.' A gentle secretarial hand has done more sums and calculated, after receipt, 'about 23/ per week'. There was a note attached:

BORVE
STORNOWAY
14th March 1919

Dear Sir

I wish to bring to the notice of your Committee the case of Wd. [Widow] Murray 43 South Shawbost who is seriously ill with influenza. Two of her children are also now recovering from the same illness. I think the case is one in which an immediate grant in money should be sent to the widow, brandy, Bengers . . . Food. I took the liberty of engaging a local woman who has been in attendance there for about a week and I think that your Committee will grant this woman something for her services.

I am,
yours faithfully
Dr R Ross

Dr Rhoderick Ross's little grandson in Yorkshire, Iain MacLeod, would one day be an eminent Conservative politician. Help was given Mrs Murray. Money was also readily granted in March 1921 to help Kenneth Nicolson, an eighteen-year-old Glasgow apprentice from 10 North Shawbost, support himself in the great city: his father, Donald – one of the oldest *Iolaire* men – had left his mother with six other young children.

One pathetic case in Point is endorsed on 4 March 1919 by Rev. George MacLeod, the United Free minister. 'I hope your Sub-Committee will consider the above family and if they think the case suitable, grant them an additional grant. The facts are: the son lost in the disaster was their only son and only support. The mother was, for a time, in the asylum at Inverness, but is at home at present able only to take care of herself, doing little or no work. The father is, I fear, following in his wife's footsteps. He also seems to be mentally defective. He is doing nothing for the support of the house. I am told that he sees and talks to no one. I found him when I called much depressed. The only other member of the family is a young girl about sixteen years of age. One feels very sorry for this lassie. How she is able to do all she does for the home is a wonder. The income is 16/6 got for their late son . . . They have no other means of any kind. I do feel that this girl, who acts almost as a slave, should get something. But that is a matter for the Sub-Committee. The above are the facts as I was told and I believe them . . .' The mother and father were but forty-nine.

The Disaster Fund evidently agonised over the sister of one Sheshader casualty, who – by then over fifty – begged from December 1932 for the continuation of the allowance paid to her recently deceased mother. 'I am sorry to inform you that I am not fit to do any work now and I have no one living to support me only one Brother and it takes all his time to keep himself alive. I have been looking forward to this allowance to keep a roof over my head and if it is not granted to me Well it would hard for me to appeal to the Public Assistance Officer for support which I hope the trustees will prevent me from doing and grant me the allowance out of the Fund I am enclosing stamped envelope for reply Respectfully yours . . .' She was sent, on the chairman's authority, an emergency grant of £3 and, later, a special payment of £5 – 'not to be looked on as forming a precedent'.

But the Committee were not as soft a touch as they might appear. The 56-year-old father of one lad from North Lochs, himself resident in Stornoway and describing himself as 'unemployed at present through ill health – Receiving pension of 11/9 per week,' put in a February 1919 claim with this rant.

Dear Sir

I am sending you this letter regarding my son who was lost in the *Iolaire* disaster.

I am his father though he was staying over at Lochs through his mother having died when he was young but I was doing to him as best as I could in keeping him and I think you don't think I am not entitled to get a portion of the subscription for the bereaved families.

He was my oldest son and I was looking forward to him as my first help and if he was spared he would do that for he was a dutiful son to me in all he could though he was giving most of his earnings to his Auntie . . . and I hope you will see through the case if you please let me know the decision. I remain yours obediently . . .

Atop the form, it is hard to fault the decisive writing: 'None'.

The crude Harris slips show general dignity. Finlay Morrison, twenty-five, would never come home to Àird an Aiseig, Scalpay; nor make peace with Ciorstag Ruairidh. His fifty-six-year-old parents, a twenty-one-year-old brother (an active fisherman) and three teenage sisters now mourned him. They had Admiralty pension of 15s and irregular income otherwise. 'I attend fishing rowing to Stockinish Head weather prevailing this last few years. I would not average more than one pound weekly,' young Donald has written. 'Allowance for mother,' the Committee decided.

Farquhar Morrison of Scrott, Stockinish, had been thirty-five. Like Finlay Morrison, he is buried at Luskentyre. He left a widowed sixty-year-old father, Donald, and had evidently been widowed himself very young; the bereft household consisted of his thirty-eight-year-old aunt (not a startling age, in a culture of very large families) who kept the place for his father and Farquhar's own eleven-year-old daughter.

Though he could not write himself – his sister or granddaughter presumably took his dictation – Donald Morrison's letters are correct and grave. 'I beg to return herewith and in connection with the "Disaster Fund" in respect of my son Farquhar Morrison who was lost on board the *Iolaire* at the Lewis Disaster. I beg to bring before the Committee the deceased here entered, the only support . . . of . . . the household. He sent regularly in addition to attendance and separation allowances of all his earnings to meet our expenses as I am unable to earn anything since a considerable number of years . . .'

The Committee seems to have been tardy with Harris claimants. Donald Morrison had to write again.

Stockinish Harris
7 August 1919
The Secretary
Iolaire Disaster Fund

Sir

I beg to apply for some assistance from the fund of the above named in respect of the loss of my son F Morrison RNR Seaman.

I receive no allowance since a considerable time in respect of him, and I was wholly dependent on him. I hope you shall consider my position and let me know by return.

Yours faithfully
Donald Morrison
X [his mark]

The Committee evidently hasted to make amends, for they received another letter on 26 February 1920.

I herewith return Receipt for £6 for which I beg to thank you.
You may note I have ceased receiving separation allowance and allotment at the end of 26 weeks from date of death of my son.

I have received expenses for myself or daughter but the child of deceased claims 10/- weekly for her own keep.

I have no certificate of birth for the child of deceased but was born in February 1908. Hope a substantial allowance will be granted me.

Yours faithfully

his

Donald X Morrison

mark

Arrangements were made for this family. This is one of the very few notes of local thanks in these records.

But not all who had lost a man, of course, on 1 January 1919 lived on the Long Island.

Another widow also completed her form: the claim made by Lucy Lavinia Mason is dated 14 February 1919. The relict of the late Commander Mason, forty-four, was only twenty-six. But she did not post it to Stornoway for some weeks, and the subsequent official scrawls on the document seem fiercer than usual. Her covering note was on the best black-bordered stationery.

75 Main Road

Handsworth

Nr. Sheffield

April 19th 19 [sic]

To Hon. Secretary

Dear Sir

Enclosed please find the form which I have filled in to the best of my knowledge. The Admiralty are to make me a pension of £200 a year.

Yours faithfully

L L Mason.

'. . . a pension of £200 a year from Admiralty . . .' dashes an incredulous pen over the form, with squiggles and scorings-out in very black ink. The birthdate of an unnamed daughter – '11/6/19' – appears on the form, but was scored out: an infant lost to the influenza, or an administrative error?

Two years later, the Widow Mason would write again.

23 Mansion Road
Sefton Park
Liverpool
July 9th 1921

To The Hon. Treasurer

Dear Sir

I have just received from you (£5) five pounds (cheque) value. I am returning same to you for change of name – or in case I am not entitled to same – having remarried last Jan 8th 1921. Will you kindly inform me at this address?

Yours faithfully
Lucy L Finlayson
nee Mason
Widow of the late Commander R G W Mason
late of 75 Main Road
Handsworth
Sheffield

The scrawl over the form is vast, twice underlined. ' . . . *Re-married . . .*' And yet Lucy Lavinia Mason or Finlayson had a life no less upended by the Beasts of Holm than any island woman and, under Disaster Fund terms, every right to its assistance.

There is plain pathos about those whom Lieutenant Leonard Edmund Cotter, forty-nine, left behind. Like so many others, they had no body for a funeral, no graveside to visit. His fifty-year-old wife, Margaret Eleanor – Cranmore House, Park Road, Cowes – found herself with two teenage daughters and a twelve-year-old son, Leonard Edgar. Her shock is evident, for she has signed two completed forms on 8 February 1919, one muddling her income and reversing her son's names, 'Edgar Leonard.' He and his sixteen-year-old sister, Ena May, were still at school. Mabel Violet, seventeen, was at 'Private Lessons'. Mrs Cotter's personal allotment and separation allowance was £16 a month; the children got £6 a month and other income for the family amounted to £55 a year, an annual income of £319.

By 16 March 1920, Margaret Eleanor Cotter was dead – perhaps of the Spanish 'flu. And her children's appointed guardians, a brace of lawyers – 'George Henry Meagen and James Charles Wilson Damant, both of Cowes, Isle of Wight' – on that date signed a new claim-form on behalf of the three children and 'Miss Wrayford' the 'housekeeper'. Meagen and Damant, on their own form, gave the total provision as 'About £300 per annum.' A Disaster Fund Committee hand has coldly written: '2nd consideration.'

Mr Damant later wrote Angus Cameron in Stornoway.

Damant and Sons
Solicitors
J.C.W. Damant

Cowes
Isle of Wight
2nd July 1921

Dear Sir,

Cotter Deceased

I am writing on behalf of the Cotter Children to ask if a further grant from the *Iolaire* fund could be made to them. The Boy is at school in Romsey and the Girls are training for Nurses.

Any further grant would be greatly appreciated.
Yours faithfully
J.C.W. Damant
Angus Cameron Esq
Hon Treasurer

It was almost certainly granted. The fate of the girls is unknown; Leonard Edgar Cotter, fifteen in 1921, was shortly shipped to a whole new life on the other side of the world.

There is a bleakness evident in other papers submitted by those who wept, from the breadth of Britain and from places of great ugliness, industrial and domestic and spiritual.

Ernest A. Brown, Trimmer Cook on the cursed steam-yacht, had been twenty-two. His body was lost in the Minch. He left a sixty-one-year-old mother, Mrs Martha Kendal, and a sixty-six-year-old stepfather, Stephen, who earned £2 a week. She had got 10s a week for her son. 'Dear Sirs,' she wrote on July 5 – probably 1919, 'With grateful thanks I received the £5 this morning my husband as [sic] had 9 weeks at home on account of the coal strike again. Yours respectfully M. Kendall.'

Charles Dewsbury, Leading Deckhand, who had enlisted from America, had been thirty-three. As we noted, he lies at Sandwick. He left a twenty-eight-year-old widow, Emily Matilda, at 1 Rodney Road in Great Yarmouth as of 8 February 1919. She had an allowance of 25s a week. Several years later she wrote a pathetic letter from Gorleston on Sea.

Dear Sir

I was informed by the person who took my house 1 Rodney Road Grt. Yarmouth that a black edged letter was addressed to me there I thought most probably it was from the "Fund." Am I entitled to the money now as I have re-married & living at the above address. I should be very thankful for it as times are very bad.

Yours truly
E Barr
nee Dewsbury
Secretary

The Committee's decision in the case we do not know.

Ernest Leggett, forty-one, Deckhand and the hapless helmsman at the moment of impact, was from Emsworth, Hampshire. He had been buried at Havant. His death left a chaotic situation: he had been a widower, with three sons and a daughter, all under twelve, and with an elderly father. 'Sir,' wrote George Legget on 12 February 1919, 'My son the late E. Leggett, Deckhand, had been living with me ever since the loss of his wife who died about 15 months ago & the 4 children were put out to be looked after as it was impossible for me to do so owing to my age 68. I am a fisherman by trade & so am not able to do much of that now. I have only one son at home now who is my sole support. But I am willing to see that children of my late son have every ease as long as I can. I remain yours etc. G Leggett.'

An appointed guardian of the little Leggets, Major J.M.K. Robinson, of the Soldiers and Sailors Help Society (Havant District) sent a 17

February 1919 form. The children had been ruthlessly separated – eleven-year-old Ernest to Lewisham School; eight-year-old Dorothy to her maternal grandmother; Fred, just six, to the All Saints Orphanage at Lewisham; and four-year-old Willie to a Mrs Giles, at Alexandra Children's Home in Stoke Devonport. George Leggett had sent his competing claim on 12 February for Iolaire Disaster Fund benefit and by March things were most involved. An Ellen Jewell becomes involved, as a mysterious 'Representative', and by 8 June 1921 the All Saints Orphanage is demanding money from Stornoway, by the agency of the Hampshire Local Committee under the War Pensions Acts 1915–1918, for young Ernest and Fred as 'I understand that Institution receives their pension of about 12/- per week. The Secretary now reports that these children, whose ages are 13 and 8 respectively, cannot be maintained for this sum . . .'

A blackhouse in Skigersta, amidst true community and with some extended family and even the meanest cash allowance, would have been a blessed state compared.

Carpenter Frederick George McCarthy, thirty-four, had lived in West Hartlepool. His was yet another lost body. He left only parents in their seventies. On 5 January 1921, a Mr John H. McCarthy – the drowned sailor's brother? – wrote Stornoway, advising of the old man's death, and returning a cheque for £5.

Henry Mariner of Portsmouth, a twenty-five-year-old deckhand, had only his parents to weep for him. They had but his £1 weekly allowance – for six months after the Beasts of Holm – and no additional income 'other than when my husband is in employment', stated Minnie Elizabeth Mariner, fifty-four. She sent a covering letter with her February 1919 application. 'I shall be most grateful for any donation the Iolaire Disaster Fund might give me. For in losing my son – I've lost my principal support for old age period. For his interest was entirely centred in his Mother & Father – & said his Dad should not work from after War was over and he'd started his Mercantile career again. There was a great love and trust between us . . . Besides which my house is made up principally from his gifts. His aim was anything and everything for the old folk's comfort. He was happy – if we were happy. Please excuse me for troubling you to read this – but my heart is so full of him – & my loss of him so great. I can't seem to help telling you. I beg to remain, Yrs faithfully M E Mariner.'

David MacDonald, Signal Boy, had been just seventeen. He, too, is buried at Sandwick. He left two younger sisters and parents in their late forties at 53 Virginia Street, Aberdeen. Jane, fifteen, earned 10s a week as a mill worker. Flora, eleven, was still at school. John MacDonald, his father, made £2 10s as a shore labourer. Their only son's 'allotment and separation allowance' was a pitiable 4s 6d a week. The form is completed and signed by his mother, Mary: she probably never saw her son's grave.

Harold Moore, Deckhand, is buried in Southend. He was survived by a twenty-one-year-old wife, May, and a twenty-three-month son and namesake. Their allowance was 30s a week. In her distress she could not quickly find her husband's official Service number. She was provided for by the Fund. Several years later, when times were hard, 'I regret having to put you to any trouble after the kindness shown to all of us but would you kindly inform me whether we are to receive a "cheque" as usual and if so the date I can rely upon the same. Thanking you all for the great help given us . . .' '£5,' a Stornoway hand has appended to her letter.

William Joseph John Stanley lies at Greenwich. He was only nineteen, a Mercantile Marine Reserve and a deckhand on the *Iolaire*. His parents and sister were finally awarded a weekly Admiralty pension of 9s 6d. On 23 February 1919, his forty-seven-year-old mother, Lucy, sent a humble note of thanks to Stornoway for the rapid gift of £5. Private Herbert W. Head's mother sent her form from Suffolk. Catherine Head was a fifty-six-year-old widow. Herbert had left an older half-brother, Alfred, thirty-eight; two brothers then demobilising from the Forces, another still serving, a married sister, and two much younger brothers who were farm labourers. Yet another brother had been killed in action in August 1916. 'Allowance for mother only' was the decision on Lewis.

And there were Albert R. Matthews's parents at Chiswick; the twenty-two-year-old Leading Victualler Assistant is buried in Islington. When his father, William, fifty-five, was in work he made £3 a week. His wife, Annie Josephine, was 'unable to work.' They got a weekly pension of 5s for their dead son. 'Allowance', the Disaster Fund has recorded on the form.

On 5 July 1921, from 27 Arthur Road, Holloway, 'Mr Cameron – Dear Sir, Many thanks for your kindness in sending to me which I feel very thankful for & will make good use of. Thanking you again, I beg to remain, yours sincerely, A Matthews.'

The aunt of 52-year-old Alfred Taylor, buried weeks earlier at Sandwick, sent a firm note.

March 7th 19

Dear Sir

I thank you for letter received this morning – I am so sorry I have given you the trouble of writing a claim. I could not fill up the form which was sent me, not being a dependent or relative to my dear boy Mr Alfred S Taylor.

My dear nephew as I have always called him from a baby was very dear to me & his death has been a great grief to me.

He had no parents. His mother and only brother died some years ago and he lost his father at 15. My dear nephew looked up me [sic] all the respect owed to his own parents. That is why he gave my present address. So the things belonging to him should be sent to me. But I have never been a dependent or anyone else of the late Alfred S Taylor.

Yours faithfully
M Allen
205 Gloucester Avenue
London W2

'Not Claiming' was written across the form in Stornoway, underlined.

John Hern, of 13 Roxburgh Street, Sunderland, father of 2nd Engineer John Hern, twenty-six, wrote emphatically on 8 February 1919. 'Dear Sir. Re *Iolaire* Disaster. I am in receipt of your form which I return as we are making no claims on your Fund. I dare say their will be more urgent cases than ours on your island. Would you kindly let me know if you are having a public Inquiry regarding the loss of the *Iolaire*?' And 'Not Claiming' is written on that form, too.

At 11 South Shawbost, on 11 February 1919, Mrs Kate Smith stared at this cold bit of paper, with all its boxes and dotted lines. She was thirty-six, her life given over to that of a wife – well, it had been – and a mother of four little children: Catherine, nine; and John, seven; and Maggie, five; and Maryann, a mite of two-and-a-half. John Smith senior, a deckhand, had served on HMS *Duchess of Devonshire*. He had been thirty-five. He had just been buried at Bragar. She would continue to receive £2 16s 4d of a weekly allowance – until 30th June 1919. The question was blunt: 'Relationship to Sailor.'

Mrs Smith thought, and wrote. 'He was my Beloved Husband.'

In Moascar War Cemetery, off the Ismaili to Cairo highway in Egypt, lie the remains of a man who fell in the vast battle of El Alamein on 24 June 1943. A shepherd in peace, he was thirty-six years old and a sergeant of the 25th Battalion of the New Zealand Artillery. It is the grave of one who gave his life in the war to defeat Nazi Germany. It is the grave of Leonard Edgar Cotter.

Chapter Eight

Dead Reckoning

Chaidh dithis de chàirdean mo mhàthar a chall air an *Iolaire*, dà bhràthair, agus bha fear an ath-dhoras, chan eil fhios agam an e sia seachdainean a bha iad pòsta – chaidh a chall. Bha an teaghlach a bha seo, 's ann air a' Ghleann a bha iad a' fuireach – chaidh an duine a bhàthadh, 's chaidh athair na mnatha aige a bhàthadh, chaidh a bhràthair a bhàthadh, 's chaidh a cliamhainn a bhàthadh. Cha robh iad pòst' ach na sia seachdainean. Chaidh mac Red a chall. Bha a phiuthar thall na choinneamh. Bha gig acasan. Bha balach eile shuas an rathad. Cha robh taigh ann an seo bho muigh an sin anns a' Chàrnan nach deach cuideigin a chall ann aig àm a' chogaidh. Chaidh bràthair dham athair a chall aig an t-Somme. Chaidh an t-aon bhràthair a bh' aig mo mhàthair a mharbhadh anns na Camshronaich – cha robh e ach bliadhna thar fhichead – anns an Fhraing . . .

Two of my mother's people were lost on the *Iolaire*, two brothers, and the man next door. I don't think it was even six weeks after he got married that he was lost. There was one family here, they were living in the Glen – the husband was drowned, his wife's father was drowned, his brother was drowned, and her son-in-law was drowned. They were only married six weeks. Red's son was lost. They had a gig, and his sister was over to meet him. Another lad up the road was lost. There wasn't a house from here up to the Carnan that didn't lose someone in the war. My father's brother was lost at the Somme. My mother's only brother was lost with the Camerons in France – he was only twenty-one . . .

Seònaid Mhurchaidh, banntrach Ruairidh Mhòchain
Mrs Jessie MacIver, 2 North Tolsta, in interview with Chris Lawson, 2008

It is Saturday, 27 September 2008 – fresh, clear, cold, with a nor'westerly wind and the odd whipping shower. After much writing, and immersed so long in such harrowing material, I rather feel like a day away from the desk. Besides, something is on my conscience, nagging at me more and more as I have made my weekly visits to Holm Point, wincing at the misleading plaque by the wicket-gate, grabbing a stone for the cairn, watching sheep scatter as I stump down the sward.

I have been here at full tide, with only the tall green beacon visible on the submerged outer Beasts. Once, too, there was a decisive southerly breeze, not a gale or anything approaching it, but which nevertheless thumped John Finlay's ledge with breakers of remarkable force and backwash, underlining just how difficult it could be for a floundering swimmer to make it ashore uninjured, or safely to make land at all. And, last Friday, there was a low, low tide, the sea at Holm now but a slack, deep, ominous swell, the abrupt fall-off from the rock exposed, the kelp and wrack a languorous forest below.

Over two hundred men died here and, even without any bent to superstition, that must somehow mark a place, perhaps in a way some future scientific generation will discover – a force somehow reverberating in landscape – and, though you knew nothing of the story, and there were no memorial at Holm Point and no recent endeavours to prettify the access for tourists, I think even the entire stranger would sense something terrible had happened there; as they say of Polish forest-groves where certain camps once stood that no bird is ever heard to sing.

I drive to Stornoway. I stop by a shop in Bayhead to buy the necessary items. At Holm, projecting and exposed, there is this day an iced cross-wind, though the sheep are now off the pasture and at least my tiny elderly dog can join me, padding in my trail, roaming in short, petulant circles once we have reached the memorial and, to her evident frustration, the expedition takes a pause.

Even in a few weeks the lettering seems more worn than ever, paint flaking off chip by tiny chip through each passing day. I step over the galvanised little fence with stiff care. I soak the rag in the white spirit. I shake the little can of black gloss. A 50p coin has just the edge and heft to remove the lid; I unwrap the little brush, stroke it a dozen times against my thigh to eliminate loose hairs, and I start.

It takes a while to touch up the lettering on an inscription such as this, and it is a long inscription. At least the words are embossed – carved out of the granite, rather than engraved into it – but I keep as little paint on my brush as possible, lest it drip; use tiny dabbing strokes, lest it spray. The top of the legend is most difficult, because I can neither stand straight nor assume a much more comfortable crouch, and within a few minutes, lopsided and stooped, my back complains: I am not now twenty-two, and will never again be young. Stroke by serif, comma by period, letter by letter, the task advances with slow, demanding concentration.

The roar of engines seems to come from nowhere. I start and turn. From the outer cape of Holm Island a squat, powerful-fishing boat surges into sight. And she does not go south and safely outside of the Beasts, but right in, within hailing distance, along Cotter's lethal course, right over the inner reef and over the grave of the *Iolaire*, at a confident nine knots and moving at all times in neat parallel to the shore. I am so shocked it takes a few seconds to realise the tide is so full she is in not the least peril, that she cannot draw more than four or five feet of water; too taken aback to wave at the shadowed figures in the wheelhouse. In half a minute she is past, clear, and lost from view into the mouth of Stornoway harbour.

It takes me a minute to renew a vital pair of quotation marks. I remember the *Sheila*, that cramped and tough little mailboat, which Captain Cameron brought so surely into port that New Year night, past the stricken *Iolaire*, the cries in the dark and the agonies of tortured iron most audible on his lonely bridge. But Cameron could not always be on duty and – precisely eight years later, on 1 January 1927 – a thoroughly deserved furlough proved fatal for his command. The *Sheila* had sailed from Stornoway late that Hogmanay evening. She never reached Kyle. The relieving master was not familiar with the passage or the many, many hazards of the west coast and the Inner Sound, between Raasay and Torridon. 'We do not know the details,' write Duckworth and Langmuir, 'but it is possible the officer of the watch never saw the South Rona light, and thinking he had run his distance, altered course with a view to making the Applecross call, and did not discover his error till too late. The result was that the unfortunate vessel ran ashore in darkness in Cuaig Bay just south of the mouth of Loch Torridon, very early on New Year's morning.'

There were no casualties – all aboard scrambled to safety over the rocks;

the surprised folk of Cuaig provided swift warmth and shelter – but the *Sheila* was a constructive total loss, and it was only the first of three 1927 disasters – the wreck of the *Chevalier*; the fire that gutted the *Gondolier*, killing her master and two of his hands – that bankrupted David MacBrayne Ltd and reduced it to a creature of subsidy, since 1969 wholly in public ownership.

The *Sheila*'s fate typifies, in drear banality, the misfortunes of Lewis and Harris between the world wars, as the *Iolaire* has come to appear an awful harbinger of death, defeat and dismay. That fraught history is well known. It can be argued – as Bill Lawson has put it to me – that the *Iolaire* put steel both in Great War survivors and in the rising male generation, just too young to serve – very many of whom sailed overseas, from 1923, on great emigration ships whose names are almost as resonant today: the *Marloch*, the *Canada*, and especially the *Metagama*. Young Lewismen, hardened by war or appalled by the Beasts of Holm, were not now in the least prepared to be denied their own land, or to knuckle down as a pliant, malleable workforce at the entire whims of a Lancashire soap-magnate. They demanded new crofts from Leverhulme, in all his imperious vision. 'How can I have that in the case of men who are in the independent position of crofters? . . . I will not *not* compromise, I *must* control . . .' he wailed to a Board of Agriculture official. 'Then,' said young Colin MacDonald, 'I am afraid you are only at the beginning of your troubles.' Leverhulme ignored his counsel, took on the men of Lewis – and lost.

And those who argue that this was a Long Island catastrophe, or have used it since to belittle the industry and foresight of island people, should remember that Viscount Leverhulme's Hebridean fantasies were despised by his own children; that the Lever Bros. board of directors had forced him to sign papers expressly detaching the whole Hebridean enterprise from the company; and that his new projects in Harris, where he had retreated, were ordered to cease at his death in May 1925, before the old man's body was even cold. Workers got but a week's notice. Leverhulme had hatched schemes to demolish and quite rebuild Stornoway that would have enchanted Albert Speer: had he won, and started, his death would have left the town rubble.

Lewis had seen off despotism: for long, it seemed she had reaped only destitution. In April 1923, thirty-five journalists and photographers converged from the mainland to watch the *Metagama* sail from Stornoway to Canada, bearing 242 young men and twenty-three island women. They all wore little maple-leaf badges and were solemnly presented with Bibles. A pipe-band played; Gaelic psalms were sung; Free Church ministers addressed the emigrants. So she sailed, passing blazing beacons on Point and by the Butt, where my 10-year-old grandmother watched the *Metagama* go by. ('And how many of the Habost boys left?' my father asked her seventy years later. She looked at him with evident exasperation. 'They *all* left . . .') One man from Tolsta, 'Noah', had given sombre orders: as the ship wallowed past the Tràigh Mhòr, friends ashore fired the thatched roof of the home he had left behind, a noted *taigh-cèilidh* for the adolescents of the district, where many happy times had been had; and those flames were his last sight of the village and his island.

In January 1924 – always the most vulnerable point in the crofter's year, with peats and supplies low and no grazing for beasts – the *Free Presbyterian Magazine* wrote of Hebridean miseries in the sort of language my generation would associate with Ethiopia.

Distress in the West of Scotland. – Owing to the exceptionally wet summer and autumn, combined with widespread unemployment, there is real distress among the people of Lewis and the Western Isles and the western seaboard of Scotland. Sir Hector Munro, Bart., Lord-Lieutenant of Ross and Cromarty, has issued an appeal to relieve this very serious destitution. The hay and the corn have not been secured, the potato crop is a failure, the fishing has not yielded good results, while owing to the wet weather, peats have not been dried. Sir Hector has made an appeal for help either in money, kind or clothing. In his appeal he says, 'It is a population left without money, fuel, or sustenance to face inclemency of the winter and coming spring . . .The crofter and the cottar are not eligible for unemployment benefit, and Parish Councils, whose rates are already over 20s in the pound, are at the end of their resources.' The Board of Agriculture has allocated a sum of £4,875, and other grants have been intimated. Our Government might employ the money sinfully spent on that useless functionary, the Envoy to the Pope (up to the end of 1922 he has cost the country £32,365), by sending him about his business, and thus allow the people of Lewis (which gave 6,000 men at the commencement of the war) to get the benefit of the sum allocated for his upkeep. We are sure our readers will feel for their fellow-Highlanders, and will not only show their sympathy in words, but in a more tangible way by sending what they can to the responsible officials who are dealing with the matter.

Sickness; overcrowding (often in poor, damp new houses of poured concrete, roofed with tin or rubberoid, in many respects inferior to the traditional dwellings); widows and orphans on all sides; wholesale unemployment; massive emigration; rampant tuberculosis – for this, a generation of Lewis people had so gallantly surged to the colours in the Great War. The public life of the Long Island reached its nadir, probably, in 1928: Town Council and District Committee affairs bereft of talent or vision, their sessions marked by bitter ineptitude. The Prohibitionists triumphed in 1921, and for six years, save in ridiculous wholesale quantities – a case of whisky, a keg of beer – a merry young man could not even buy a drink.

There are tales, certainly, of strife and spite. Desperation for a croft – which was, at least, a life-raft against entire starvation – cannot be underestimated as a factor in the Lewis of the 1920s. The death of a patriarch could (and did) see widowed daughters-in-law and their infants thrown out by a returning heir-apparent from Glasgow, and often with the desolate grandmother to boot. And the hunger for land was, if anything, eclipsed by a still more consuming greed for money: anyone with a cash income, even with a Great War pension – and especially an *Iolaire* widow – was tacitly resented.

But all this reflected, of course, national political failure. Between 1918 and 1925 there were no fewer than five general elections and a sustained instability as Labour edged the divided, declining, fractious Liberals from major-party eminence. Whoever was in power seemed to make very little difference. Britain was saddled with vast war debt to America, a falling birthrate, an economy still geared to outdated heavy industry rather than the manufacture of new, light consumer goods people actually wanted to buy. Between 1921 and 1939, unemployment in the United Kingdom never fell below 10 per cent. And, from 1931, when the well-meaning King George V banged heads together to engineer a 'National Government', the country had no effective Opposition. Britain thus sleepwalked through credulity and appeasement to the Second World War.

On a markedly reduced population, Long Island commitment to that struggle eclipsed, if anything, her part in the First. From the 25,205 population recorded in the 1931 census, it has been reliably assessed that at least 5,500 sons of Lewis engaged in the struggle to destroy the Third Reich, incurring heavy losses – especially – in the Battle of the Atlantic, not least in that forgotten Cinderella of HM Forces, the Merchant Navy. North Tolsta is said to have had more men in this war, as a proportion of her people, than any other community in Britain: twenty-three of them would be killed.

Incredible as it may seem, at least two *Iolaire* survivors – Neil Nicolson and Alick Campbell – served at sea throughout World War II; indeed, Campbell was one of fifteen Harrismen to fight in both world wars, and the Harris endeavours against Hitler (as recorded in a moving, 1992 *Roll of Honour* volume) probably exceeded that of Lewis. The isolated hamlet of Molinginish, on the southern shore of Loch Seaforth, paid the highest price. Of her five young men serving, two were lost at sea; a third spent five years as a prisoner in Germany. Within a few years of peace, Molinginish was simply abandoned. No one now lives by the music of her waterfall.

Others who had escaped the Beasts were past their fighting days. In 1942, matching the mood of a nation newly martial, appeared *My Amazing Adventure: A Collection of True Stories*; and amidst assorted ripping yarns we find 'Wrecked on the Beasts of Holm' by "L. Welch . . . now living at Malvern and . . . employed as a commercial traveller'.

Two decades on, Welch's generously embroidered account varies in significant detail from his Admiralty Court of Inquiry testimony, and is not without entertainment value. 'When the crash came I was preparing to go off watch. For a second the ship seemed to be lifted out of the sea and held suspended in the air. Then came a shattering impact and I was flung across the wireless cabin . . .' Between a 'floor that was almost vertical', water that was 'rushing in' and the odd 'crazy lurch' the cabin was 'plunged into darkness. A cable had gone.'

Welch fought his way outside, into a 'black fury of wind and driving sleet. As I made my way on deck I heard the confused jabber of voices and shouting in a foreign-sounding tongue, and remembered that the three hundred passengers aboard HMS [sic] *Iolaire* were Gaelic-speaking men from the Outer Hebrides.'

Telegraphist Welch rallied bravely from this appalling discovery. 'They lay in tangled heaps on the deck. As I staggered and clung, I saw great breakers rolling over the ship with a crash, blotting out these dark figures. The next instant the waters seethed and whitened and washed them off like flies . . . Still uppermost in my mind was the wireless transmitter. I had to get it going somehow. A sailor lurching by shouted that we were on the reef. Lashed by wind and sleet, I made my way to the engine-room where with the help of the chief engineer I mended the broken cable to the wireless cabin.'

Back at his sparks, he was briefly joined by Commander Mason 'who calmly gave me the ship's position . . . Before I could begin transmission the ship gave another terrific lurch. I had a sight of the square-faced Commander as he was turning to leave the cabin. Then there was a splintering crash and the lights went out again. Almost immediately I was waist-high in water . . .'

Welch diagnosed a failed bulkhead, and amidst a welter of water, lurches and adverbs not only got back to his cabin and secured a life-jacket but an electric torch besides, before trying 'to fix up some emergency transmitting gear. The Commander had gone. I never saw him alive again . . . I struggled vainly to get the wireless apparatus going, but rising water forced me to abandon the attempt. How long it took me to get on deck again I do not know. Throughout the wild adventure I completely lost track of time. All I remember was the howling wind, the blinding red-hot sleet and rain, the struggle to reach some object to which I could cling a few feet ahead . . .'

Welch, by his own account, was almost swept away by an 'enormous sea', clung 'half-conscious' to a stanchion, hailed that 'hefty, tough American' Charlie Dewsbury who was working at rockets – " 'Ruddy fine night for fireworks!" I shouted' – and somehow, amidst all this drama, reached the bridge. There Lieutenant Cotter 'had lashed himself to the rail, determined to stick it out to the end. Giant seas were breaking over him . . . "It's abandon ship," he yelled. "Carry on." "What about you, sir?" I asked. "I'm staying here," he replied.'

Welch was about to go over the side when he remembered he had no money. He returned to the cabin to pocket a few notes – this is perhaps the most incredible detail of his narrative, capped only by the next: noticing a light still burning in the galley, he forced his way in and 'sat on a locker taking stock of my chances of getting ashore . . . I took stock of the beautifully appointed galley of the ship and thought of her past . . . She still retained traces of her pre-war luxury, although most of her beautifully polished panelling had been boarded up. In my wireless cabin was silk tapestry, and these floating cooking utensils reminded me of her peaceful past, stressed the tragedy of her present plight.'

Happily for posterity, Welch decided to abandon these meditations on naval architecture and to make his escape: by this time the door had jammed and he had to smash a galley window. From this point there is a clear ring of authenticity and no further contradiction of his 1919 statements. 'On the main deck I made my way to the side and selecting a dark patch of water, plunged in automatically. Then began a struggle of life. It was wild, blind, instinctive. I knew nothing, felt nothing. I could only strike out until I was exhausted, then abandon myself to the fury of the sea. My sense of time was suspended. I had no idea of direction . . . Once I collided with a dark object that must have been a body. I felt no sense of horror, only that fixed determination to avoid being smashed to bits on the rocks . . .'

The telegraphist made it to land, helped another survivor on – 'locked arm-in-arm we stumbled forward together in silence, broken by sobs and gasps' and to the shelter of a 'crofter's cottage' – not a description the Anderson Youngs might have appreciated. 'Within an hour I was in the naval barracks at Stornoway, between thick blankets . . . Only 30 survived the tragedy. There were some miraculous escapes. A boy named John MacDonald climbed up the mast, lashed himself there, and was taken off at 10 o'clock the next morning. As soon as I got up, I went into

Stornoway to buy some shaving kit. The shopkeeper was an old woman with red cheeks and pale blue eyes. As I tendered one of the half-sodden notes I had stuffed into my life-jacket, she looked at me in superstitious horror and shook her head. "If the Lord God pulled you out of that wreck," she whispered, "I couldn't take it." '

In fairness to Welch, this colourful tale – with its contemptuous reference to Hebridean passengers, the contrasting heroism of English officers and not the least thought as to how the ship had been wrecked in the first place, is firmly from an age of wall-to-wall propaganda and has more than a hint of 'as told to . . .' journalism: its purple prose and unlikely detail certainly suggest the work of another imagination. A more sober account survives: the transcription of a talk Welch gave the British Legion in Stornoway, undated but probably pre-war.

It is important not to suggest the lull between one global conflict and another was a dark, unrelieved litany of Long Island woe. The aftermath of the Great War did bring new, if still very limited, opportunities for her womenfolk – the vote, of course; and growing employment prospects. Two daughters of the *Iolaire*, at least, proceeded to higher education. While veterans left the country, most emigrants had been too young to fight. Certainly some never returned. Most suffered from dreadful homesickness – so anguished, in some cases, they could never even bring themselves to write home, and simply vanished from island families, community and consciousness. Some – such as a great-grand-uncle of my own – only visited Lewis after Hitler's war; he, Roderick Thomson, 3 Habost, had been away so long he had quite lost the ability to speak Gaelic. Some emigrants, by hard work and sacrifice, built good lives in Canada. Not a few even found subversive employment in the bootlegging runs of the Great Lakes. Most struggled. Dozens – especially in the terrible years of the Great Depression – disappeared entirely, starving to death by the dustbins of frozen cities, consigned to the pauper's grave.

But – toughened by war, by the *Iolaire* – young men at home now rebelled against arduous hard-scrabble agriculture; resisted at last, too, the controlling patriarchy within their families and their townships, the austere and mean fathers who expected Great War heroes still to knuckle under, who were determined nothing in their world should change. Things did, despite an ageing generation, begin to change now; and that for the better. And this radicalised generation – who, to widespread surprise, in 1935 elected the first Western Isles Labour MP – fought through and beyond another wretched war to win a better world. Far from being repelled by the *Iolaire* debacle, more had signed up between the wars for the Royal Naval Reserve than ever. There were assorted motives – the useful training, the jolly camps, a chance to see the world and widen one's horizons, a genuine and even jingoistic patriotism – but it would be foolish to deny economic pressure. The annual bounty, modest as it was, bridged that slender margin between content and misery.

And many island emigrants – not a few now married with children – had actually sailed back home by the time of Hitler's war, preferring Hebridean subsistence to American hunger. These children – now in their seventies and eighties – still hold Canadian or American nationality; and the names of United States forces adorn some modern Lewis war memorials. And, for all the darker outbreaks of human nature, there was much selflessness, too, as Mary Crane – born in Point in 1910 – would relate to Calum Ferguson:

> The next of kin of the men lost on the *Iolaire* each got a gift of money. War widows, the likes of my mother, got only a small pension – more than five shillings a week, I think. There was so much money put out for each and every child, and there were six of us. Some of the families who had lost their breadwinners and their *bata làidir* [mainstay] were quite poorly off. I remember my mother sending basins full of meal this way and that in the twilight, so that people wouldn't see that she was sending those gifts to the poor. I was usually the one carrying one basin and Ewen was carrying the other – oatmeal and flour. As a matter of fact, I would go as far as saying that it wasn't only our own family that my mother's widow's pension brought up. In those days, most people in our community – but not all – behaved as though we were the one family. They were kind to one another . . .

As an *Iolaire* child in Lionel, Peggy Murray was well aware of her mother's straits. They had no croft of their own – no corn or potatoes, no milk, no eggs; and the little Admiralty pension, with the irregular subventions of the Disaster Fund, minimised charity from those about them. Dolina Murray might toil all day for neighbours on this or that chore of the season and – unlike others lending like assistance – be rewarded at evening's end with one scant bucket of potatoes; a creel of peats. Survival depended on guile, resourcefulness, the most careful budgeting.

'We had to buy everything,' Peggy remembered seventy years later, wry pride still evident. 'Everything depended on how she used the pension – we weren't well off or anything like that, but we didn't starve . . .' And the twentieth century continued to batter on this Mrs Murray, doughty little widow of the *Iolaire*, in black from head to toe since the stuttering boy had trumpeted her loss. One beloved son contracted tuberculosis of the spine, suffering so greatly that Dolina Murray found herself praying the Lord would let him die. Her prayer was granted. In renewed war, she would lose another lad, torpedoed off Malta. Yet, when two little old sisters in the district grew too feeble to walk to Free Presbyterian services, Mrs Murray – herself Free Church – simply took them in under her own roof, much nearer their place of worship, and cared for them as her own kin.

From the darkness of the 1920s a certain humour – even a gaiety – began to renew island life in the 1930s. Villages organised concerts, dances, keenly fought badminton tournaments. The Lewis and Harris League was established, and the exuberant summer-only football season has been a proud island diversion ever since. The island church was refreshed by times of remarkable revival. And one Tolsta man, George Morrison, a gifted scholar and a hilarious wit, assumed a humorous column in the *Stornoway Gazette*, and for half a century the genial satire of 'The Breve' – laced with delicious light verse in both Gaelic and English – enchanted Lewis people all over the world as he mused on the misadventures of his community, the inadequacies of her leadership and the distinctive ploys of island life. 'The many exiles from Glen Tolsta will be pleased to hear that a road is to be constructed to the hamlet,' purred the Breve. 'Some day they will be able to walk there and see where folk used to live before there was a road when they no longer live there . . .'

In the 1950s, when one 1898 MacBrayne veteran, the *Hebrides*, still chugged up the West Highland seaboard, the Breve could reassure his public that despite her venerable age there was 'no truth in the rumour that she took part in the Battle of Lepanto, nor was Vasco da Gama ever captain of her.' Morrison feared solemnly for the consequences of traditional, covert island courtship when Lewis was at last blessed with electric light. 'Crafty bodachs and cailleachs will now have a private switch by the *bòrd slios* [a plank at the end of a box-bed in the blackhouse to contain the heather mattressing] so that as nimble-footed suitors trip quietly past to *leabaidh Sìne 'Ain Bhàin* [Jean daughter of blond John's bed], the room will suddenly be floodlit, the dog will see where to bite, the bodach where to aim and the cailleach whom to criticise.'

Nor was the Breve impressed by the staid, slow, infrequent Gaelic radio and its extraordinary neologisms. 'As we go to press, we gather from the BBC Gaelic news that Hitler has invaded Czechoslovakia and that Mafeking has been relieved. We often think as we listen in the afternoon that we are hearing yesterday's news in tomorrow's language. Very *tric chan eil mi fhèin ga chomprehendigeadh* except when gems such as these might be vouchsafed – *Tha na doctairean ag iarraidh air an TUC establishigeadh Commission airson riaghailtean ùra do na Unions mu thimcheall containerisation . . .*' ['Very often I don't understand except when gems such as these might be vouchsafed – Doctors want the TUC to establish a Commission to make rules for the Unions about containerisation.']

But one New Year could never be the stuff of jokes.

About 1932, Peggy Murray left for college to embark on teacher training. A bright girl, hardworking, vivacious, she had already won a bursary to attend the Nicolson Institute. Now, this day of adventure, she left her native island for the very first time, 'and we were sailing on a lovely summer day, sailing out of Loch Stornoway, and I saw a big post there, tall above the sea. I was thinking how red it was. I said, "What's that tall post above the waves?" and there was somebody who was going over with us – we were high on deck – and they said, "That's the spot where the *Iolaire* went down."'

Peggy knew, of course, about the *Iolaire*; could vaguely remember the loss of her father. But she had never heard any details of that night; had never been able to discuss it with anyone. She stood rooted to the deck as the *Lochness* sailed on, as other girls giggled and ran about, and she thought of her father and mother, and stared and stared until the beacon had long vanished in their wake.

Peggy Murray returned to Ness as a teacher at Lionel Public School, where, of a Friday evening in peace and war, staff would set up a gramophone in the gymnasium for joyous dances, 'tripping the light fantastic', as she cheerfully recalled in our own century, to the strains of Jimmy Shand, Glenn Miller, Dick Haymes. But she met someone: great changes came. By 1955, she was a Christian, the wife of the Rev. Donald Gillies, the lady of a Free Church manse – and her mother, now sixty-four, still swathed in neatest black, came proudly to live with them, rejoicing in all the life of the home and the Free Church, and her growing, doting grandchildren. 'You know, my dear,' she declared to Peggy, near the end of her life, 'the Lord has been very good to me . . .'

Yet we might note that, as some fled the island of the *Iolaire*, one or two strangers on the margins of the tragedy were oddly drawn to it. Thomas Gusterson, the naval diver, would never forget the entangled, blundering dead about him as he dove obediently to the wreck on 2 January 1919. But in time he began to court a local girl. They married, and made their home on the island, and his descendants are on Lewis still.

Dolina Murray, widow of John, long drowned on the *Iolaire*, died in February 1967 after a short illness. She was seventy-five. In 2005 Peggy Murray still shuddered to recall that jolt of seeing her mother laid out for burial, tiny, at peace, so strange in radiant white. But her baby brother – only an infant when the *Iolaire* orphaned them all – broke down completely. ' 'S e tha a' toirt gal ormsa – carson nach do dhreasaig sinn i ann an aodach soilleir aon uair, ach am faiceadh sinn cò ris a bhiodh i coltach?' he wept. 'It's making me cry – why didn't we dress her up in bright clothes just once, so we could see what she would look like?'

In 1948, amidst peace renewed and under a radical, reforming Labour government, a ring-netter from Campbeltown, the *Amelthaea*, sailed into Stornoway harbour. Neil McArthur, a 23-year-old hand on the latter, was eager to go ashore, but his grandfather – the boat's skipper – flatly refused to join him. Back in Stornoway for the first time since 1919, James MacLean could simply not face the town: though, as if to remember a better day, he had named this craft after the original name of his old steam-yacht in her pre-Lewis career.

Then a familiar face appeared, beaming, on the quayside – 'a wartime friend – a Lewisman by the name of MacKenzie, who used to visit the MacLeans when in Campbeltown . . .' He went aboard and, after quiet conversation, took the grizzled old fisherman into town. Was this in fact Norman MacIver from Arnol, the young rating whom, that New Year morning, MacLean had coaxed to safety by the line of a grounded lifeboat?

An hour after I began, I am done. I stand back to inspect what is now most clear and contrasting elegy.

ERECTED BY
THE PEOPLE OF LEWIS AND FRIENDS
IN GRATEFUL MEMORY OF THE
BRAVE MEN OF THE ROYAL NAVY
WHO LOST THEIR LIVES
IN THE "IOLAIRE" DISASTER AT THE
BEASTS OF HOLM
ON THE 1ST OF JANUARY 1919
OF THE 205 PERSONS LOST,
175 WERE NATIVES OF THE ISLAND
AND FOR THEM AND THEIR COMRADES
LEWIS STILL MOURNS
WITH GRATITUDE FOR THEIR SERVICE
AND IN SORROW FOR THEIR LOSS

"DO CHEUMA THA 'S AN DOIMHNEACHD MHÒIR
DO SHLIGHE THA 'S A' CHUAN:
ACH LUIRG DO CHOS CHAN AITHNICH SINN,
THA SUD AM FOLACH UAINN."

SAILM LXXVII 19.

'Thy way is in the sea, and thy path in the great waters, and thy footsteps are not known . . .' All over Lewis, that 1919, ministers and men, women and boys and girls, numb with grief, had thus tried to weigh the incomprehensible in terms of the order of their world, the tenets of their faith and the yearnings of their hearts. This text, indeed, was chosen by the Free Presbyterian minister of Stornoway, Neil MacIntyre, as he ascended his pulpit stairs on Sabbath, 5 January, to face a stunned people. His Free Church colleague in town had been given a mind to expound 'Be still, and know that I am God,' and together pastors and people had stared into mysteries hard, unfathomable, and by faith endured.

The confusion over *Iolaire* casualty numbers is evident; perhaps, rather, how in truth to define a 'native of the island'. The airbrushing from the narrative of Harrismen and mainland crew still shames us. But even this dignified obelisk, with no symbol save the enroped anchor of the Royal Naval Reserve, was not erected until 1960 and, indeed, was a matter of controversy, resented by not a few. It was put up largely on the initiative of a North Tolsta man, Allan Cameron, and with little general support. The Great War, HMY *Iolaire* and her last voyage had become something of a local complex.

When last I renewed these words with paintbrush and pot, in August 1988, three *Iolaire* survivors still lived. The monument abides. The records, the press clippings, the memoirs, the BBC recordings, and the traditions handed down in townships and through families – these endure.

But even Neil Nicolson, last of all who boarded at Kyle on that New Year's Eve, indefatigable and sturdy, went finally to his rest on 21 June 1992. That year, too, Flora Boyd – whose betrothed, John MacAskill, had been cast up drowned at his own Sandwick door – passed away quietly. She had devoted here life to nursing and, in 1948, had been honoured with the British Empire Medal. Flora never married; she kept John's picture by her bedside to the end. And on 28 July 2006, just two summers ago, Mrs Christina Morrison died peacefully on Scalpay. Born in October 1895, she was the second-oldest person in Scotland and certainly the oldest ever to have lived in the Western Isles, alert and independent until the last months of her life. The chief mourner, days later in the flower-jewelled beauty of Luskentyre, was her eighty-seven-year-old daughter, and there *Ciorstag Ruairidh* was laid by her late husband, several of her children and – in sight, yards away, up the slope, the lonely grave of Finlay Morrison, drowned by the Beasts of Holm. Thus the last heart touched most intimately by the *Iolaire* beat no more. And, in these Outer Hebrides, all the adults of the Great War finally slept in history.

Peggy Murray passed away in March 2006. And of the handful of *Iolaire* orphans now living, this autumn of 2008, only one can clearly remember that January horror, Mrs Marion MacLeod of Brue on the west side of Lewis.

Mòr Bhrù, as all know her, was born in 1914 as Marion Smith of Earshader, daughter of Kenneth, lost on the *Iolaire*. A second cousin of my mother, she has lived a full, most active life. A smart, inquiring girl, she spent the Second World War as district nurse for Barvas, Arnol and Brue, covering a considerable area by bicycle and providing primary health care in a day when a fee had to be paid for consulting a doctor. Mòr delivered just about every baby born in these townships for a decade – all invariably, in those days, home confinements; and is thankful she never lost either infant or mother. In 1947, she married a local man and, as was then obligatory, retired to keep house, raise a family. In latter decades, her mind stocked with lore and genealogies, poetry and song and folk-medicine, she has become one of our most eminent Lewis tradition-bearers, widely consulted and a frequent guest on Gaelic radio and television. As recently as August 2008, she travelled to Shawbost for a formal conference on the Outer Hebrides and their part in Britain's wars.

But weeks from her ninety-fourth birthday, sitting in state by a great pile of books, magazines and newspapers, Mòr sighs that old age is starting to catch up with her – she cannot now read for long; nor walk for any distance, though she still attends church and takes keen interest in everything and everybody. There is nothing senescent in these bright blue eyes, nor weary in the cultured, clever voice.

I raise the *Iolaire* most diffidently and there is the slightest of pauses – as always, in any civilised Lewis gathering.

Mòr has related her 1919 memories for decades: the clothes airing before the fire, and the brazen neighbour removing them; the funeral, days later, and more people massed in and about the house than the little girl had ever seen. But what had her mother, in after years, told of that night?

'We never spoke of it,' says Mòr calmly. 'I never once asked her.'

My incredulity must be evident, but it is true. 'She was at home most of her life, in Earshader, looking after Kenneth, my brother. She spoiled him, really. He never married, of course, and he was bad for going out, ceilidhing, leaving her alone.' (Kenneth J. Smith was a bard, a mystic, a noted Lewis character, whom Mòr at once loved and found most impractical, exasperating.) 'He was just so used to her taking care of him that he forgot, really, it was time for somebody to look after her. Well, when she was eighty-nine, John and I took her down here. That was three years before she died. What a difference it made! One of us would always be with her – one at home and one at sermon, and so on. If we were working down the croft, one of us would go back every hour to check she was all right. We would get visitors in to see her, and once a year or so there would be – oh, a 'kitchen-meeting'; they would keep a service here in the house for her, and it would be full of people, and she loved that. She just thrived.'

Had the old lady all her faculties? 'To the very end. She wasn't even ill long – less than a week. It was November 1980 when she passed away, and she was ninety-two years old. And she was the last – the very last – of the widows of the *Iolaire*.'

In November 1980, Mòr herself was nearly sixty-six. And they never talked of it?

'We never once asked her about it, and we never brought it up in front of her, and the children knew it was never to be mentioned when she was in the room.' There is not a hint of bitterness, regret or embarrassment. It was simply the tacit rule as Marion Smith grew up, one she respected to the very end.

Mòr rises carefully from her chair, grabs a favourite stick, negotiates the room. 'Kenneth Norman – my younger grandson next door – is twenty-one next weekend, and I will show you what I am giving him. It is something my father had . . .'

A capable hand collects some items from the gatefold table by the wall and passes them to me. One is a little brass box, attractive, its embossed lid centred with a royal silhouette. It is a 'Princess Mary' Christmas gift-box, sent to all members of the Forces in December 1914, filled – for most recipients, save boys and women – with tobacco, cigarettes, a little lighter. These were financed by public appeal. Inside, now, there is a folded gift of money, and some little cards. They are ration cards – cards for a sailor on leave – of the late Great War; coupons to be snipped off for tea, cheese, butter, bacon, sugar, meat or a 'meat-meal'. One has been endorsed by an officer, confirming that Seaman Kenneth Smith has been granted leave until January 14, 1919.

There are rust-marks where staples once bit; some hint of water-staining. 'Was this . . .?'

'Yes, he had it with him. It was on the *Iolaire*.'

Of late I have stood, again and again, on Holm Head; listened to recordings of long-dead survivors, visited graves all over the island. I have gone through the Disaster Fund papers, tried on Donald MacIver's lifebelt – he kept it all his days; his family donated it to the Stornoway museum – watched divers' footage of the corroding remains, touched fragments of the ship herself, brooded so intensely over that night I have scented steam and hot oil in my nostrils, had uneasy dreams of drowning. But this in my hand, these faded little cards, brings it all intimately – lurchingly – home.

I have something to add to these little things of family history – a photocopy of Mrs Christine Smith's form of application for the Iolaire Disaster Fund, and her neatly written little letter of 1920. Were this bad reality TV, Mòr would now be overwhelmed: respond with a gasp, emotion, tears, a hand over her mouth. She has never before seen these documents. But she accepts them with easy grace, dons her glasses, looks at them for a brief, inscrutable moment. Then she puts her spectacles away and gazes at me. I know, without it being said, Mòr will study them in her own privacy when I am gone; and if there is a dam of emotion to break (which I rather doubt), it will flood only then. 'Thank you. I shall give these to Kenneth Norman, too.'

I ask my wise old kinswoman just how significant – how dreadful – an event the loss of the *Iolaire* was.

Mòr shudders, sits a little straighter. 'It was the – the – the most *awful* thing you can imagine. All those men. Men of your own age, younger, the working men, those who supported families. The breadwinners, we would say now. And all at once. It was at New Year, and the worst thing – those women, those wives and mothers, they had been worrying so long, all through the War, wondering if their men would be safe and would come home. And they had thought, now, their worries were over, because the War was finished, and then . . .'

We sit in silence for a little.

'It was so strange how it fell out for some villages,' I say. 'There was not a man from Tong on that boat. There were three men from Cross, and they all survived; and all the men from Cromore survived. But all the men from Sheshader drowned. There was no one aboard from Callanish . . .'

'There was no one from *here*,' says Mòr, 'from Brue. But there were six men from Crowlista – a small, small place – and they were all lost. There was a woman from Crowlista, and she died only last year, and that girl's mother lost her husband and she lost her brother, on the *Iolaire*.'

She gathers her thoughts.

'You know, it wasn't until the second day – not New Year's Day, but the day after – that word came to Earshader, nearly two days after it happened. There were no telephones then, and we were five miles from the nearest road. The motorboat, MacRae's boat, stopped at New Year's Day with the mail as usual, and the boy from Kneep was on her, the survivor, John MacLennan, and he didn't let on – he didn't come out, he would not tell anyone.'

'He would not even tell his mother,' I say, 'and when the neighbours came, men to ask about their sons, he couldn't face them; he hid away.'

'It was hard for these boys,' says Mòr, 'hard for those who had survived. They were under such pressure. Of course, word did come, and the house was full of people, but I always went out when they came, because there would be weeping, you know; the women were crying.'

'That would be quite frightening for a little girl.'

'It was very disturbing, yes. But my mother – she carried on, you know; she did not take to her bed. There were others. Do you remember John MacLeod, the minister from Arnol, that was in Oban?'

'He was a great man,' I say, 'a fine Christian, a singer. I met him once. He composed songs. He did a Gaelic song about the Falklands War.'

'Well, his father was lost on the *Iolaire*, of course. And his mother – she was a Christian, but her mind went. It just broke.'

A memory moves. 'She used to wash the uniform . . .' I say.

'And I remember it. Every year she took out her husband's uniform, and she washed it and washed it, and she would dry it outside on the wall.'

The day is still bright, fresh, but something of horror has stolen into the room. Broken, half-deranged, Mrs Norman MacLeod of 13 Arnol spent far more than she could have possibly afforded on a mighty gravestone for her man, complete with inset white-marble plaque and the ship's name spelled wrong. She died in 1933, still only forty-nine, and is – rather pathetically – buried on the other side, the modern side, of the Bragar cemetery. Like Dolina Murray, and several other *Iolaire* widows, there was no space left for her by her husband's grave.

Mòr looks at me. 'It was the strangest thing, how that boat went on the rocks. How could it happen? What do you think happened?'

Through the long, long years of silence, the *Iolaire* festered at the back of Lewis minds, a name only overheard in hushed, grown-up talk. 'Chaidh a chall air an Iolaire,' they would say suddenly of someone recalled in discussion, 'bha e air bòrd an Iolaire,' and the very word was a game-changer, begetting awkwardness and a change of subject, as if none could bear reminding who had been aboard, who had been lost, as if the ship herself belonged to the worst of nightmares.

In 1956, though, what purported to be one survivor's account appeared in a Canadian magazine. It was lifted and carried, months later, in the *Stornoway Gazette*, where it would inspire Iain Crichton Smith to write what remains the only great English poem on the disaster. The vivid essay can still be read on one *Iolaire* website. Ironically, there are good grounds for viewing it with profound suspicion – its purple passages, the first fables of a 'wooden boat', roaring fires, the screams of burning sailors. It seems much more a piece of cynical fiction than a description by a man who was actually there. Yet – as if provoked – tongues were at last loosed. Fred Macaulay's outstanding Gaelic documentary was broadcast.

Many survivors had granted him interviews. More programmes followed, and those sturdy old men continued to speak of that night, in Gaelic and in English. The monument went up on Holm Point.

In September 1970, men diving for clams off Holm poked among the wreckage, and recovered two significant artefacts – the builders' plate from the yacht's engines, and the very bell of the *Iolaire*. Cleaned and polished, they were formally presented to Stornoway Town Council at a service in June 1971, attended by several survivors and the children and relatives of many other passengers.

'My brother and myself and my husband went to the Town Hall,' Peggy Murray recalled. 'Those that lost dear ones on the *Iolaire* were attached to one another. They found each other throughout the island . . . Anyway, there was the bell, and I looked at it, and I remembered my mother's dream.'

A television documentary was screened in 1974. The first measured analysis of the disaster, in excellent Gaelic and drawing heavily both on the relevant public records and the testimony of the dwindling band of survivors, *Call na h-Iolaire*, would appear in 1978, following successful re-publication in 1972 of a *Stornoway Gazette* booklet, *Sea Sorrow*, consisting largely of the paper's 1919 reports and some sober additional matter by James Shaw Grant.

But the *Iolaire* – not least after decades of being discussed furtively, if at all – had by now already begotten her own folklore; launched her own enduring myths.

The most fantastic merits no consideration – that her officers deliberately ran her aground. 'I have heard it asserted, on more than one occasion,' wrote James Shaw Grant in his 1982 collection of *Stornoway Gazette* columns, *In Search of Lewis*, 'that there were high words between the Captain and some of his Lewis passengers at Kyle before the vessel sailed, and that the Captain was heard to threaten, "I'll dip their heels before they get to port!"' One of his 'informants', Grant asserted, was an elderly Raasay crofter who claimed to have witnessed an altercation at Kyle between Commander Mason and some of the Lewis RNR ratings – 'in a very allusive, periphrastic manner, he tried to create the impression that the Captain was determined from the start to sink his ship.' But there is not the least 1919 testimony that Mason was anything but amiable as the libertymen came aboard, even if we could accept the fantastic scenario that he deliberately wrecked his own vessel, forfeiting indeed his own life.

Much more persistent, and at least believable, is that one or more passengers hurried to the bridge to warn the officers they were off course and the *Iolaire* was in mortal danger. Various versions of this are still told in Lewis – in a popular variant, Cotter (or Mason) pulled a gun and ordered the helpless reservist out, even as the yacht piled blithely on to destruction. But no such story was told either at the Court of Inquiry or at the subsequent public hearing. Several survivors, indeed, insisted under questioning there had been no time to get to the bridge, even if – as Dòmhnallach quite properly points out – anyone under Royal Navy discipline would lightly have challenged the seamanship of an officer. In fact, the tale never seems to involve a survivor at all – it is always told of someone else, someone who drowned, and he always seems to hail from the locality where the legend is told: in Calum Ferguson's version, it is a Point man; on the West Side, the bold bridge-visitor came from Shawbost, and so on.

One story that a survivor did relate many years later – Dòmhnallach's Gaelic narrative is confusing, but it seems to be Donald Morrison, Am Patch – does disturb. Ascending briefly to the bridge with a friend, the Patch asked if they could take two lifebelts which lay at the side. But the Captain – it could have been Cotter – said emphatically, 'Don't dare to touch any of them.' The officer had a 'revolver' in his hand, but did not fire it.

Morrison may well have related this in early questioning for the Court of Inquiry: he was not among those cited to give formal evidence, proof (admittedly thin) both that the Admiralty found his story embarrassing and, besides, he had not really been rescued by Lieutenant Wenlock. Rather than a 'revolver', it is much more likely the officer toted a 'Very pistol' – probably a flare-gun manufactured by Webley and Scott, standard Navy issue in the Great War. Some witnesses certainly insisted they saw Cotter firing one.

In any event, so inadequate were these poorly designed lifebelts and jackets that possession did not make the least different to the odds of *Iolaire* survival. Though each had his own lifebelt, the proportion of surviving crewmen was not the least higher than the proportion of surviving passengers. Indeed, Tarbert man John MacKinnon was almost drowned by his, which came loose as he fought his way back from a swamped lifeboat, with the result that 'my feet were coming up and my head going down.' That kept so tenderly by Donald MacIver, and which I was allowed this autumn briefly to don, would certainly not have kept me afloat. The story does have the ring of truth and may be the origin of the warning-to-the-bridge myth; it does not make us like Commander Mason. And might it bolster a taller tale?

For we remember Murdo MacFarlane's insistence he had seen the captain's body, with the two incriminatory lifebelts. That said, it is most doubtful MacFarlane – then a boy of seventeen – had ever once seen Commander Mason in life, or that matter many dead bodies at all; nor, for the uninitiated, is it so easy to tell one rank of Naval officer from another, merely from jacket detail.

There is certainly no official record of recovery; Mason has no grave. We would be sailing on the far frontiers of paranoia, really, to suggest the body, so embarrassingly clad, was secretly disposed of – or deliberately disfigured. Yet the yarn of Mason's lifebelts is enshrined in Stornoway thought. Jacky Morrison, a highly respected local skipper, writing to James Shaw Grant in 1981 to denounce the Admiralty files as a 'whitewash', insisted: 'No mention was made in the Admiralty Inquiry of the body of the Captain of the *Iolaire* being found on Sandwick Beach wearing two lifebelts. There are reliable witnesses still living who can testify to this shameful fact . . .' Presumably MacFarlane, who died the following year, was one of them.

James Shaw Grant would also assert, in 1983, that the wreck of the *Iolaire* was covertly salvaged – for scrap – within a decade of the disaster. 'My Vancouver correspondent, Iain Young, commenting on the fact that some years ago, when divers went down, they found little trace of the wreck, apart from the nameplate from the boiler [sic], offered an explanation. In 1927, or 1928, when a firm of scrap merchants bought a number of wrecks around Stornoway harbour, they broke up the *Iolaire* as well . . . Today there would be an outcry if anyone proposed to interfere with a wreck surrounded by so many sensitive memories. Why was there no row in 1928? . . . the public life of the island was at just about its lowest ebb. We were at the bottom of the trough from which we are now clambering out. And it may be that the operation at the Beasts of Holm, where the wreck lay, went on surreptitiously under the cover of a larger salvage operation.'

Such a suggestion is derided by J.J. MacLennan, Chief Executive of Stornoway Port Authority. The *Iolaire* is, in fact, nowhere mentioned in the records of the former Pier and Harbour Commission, even in January 1919; for one thing, movements of any Navy vessel in wartime were classified; for another, the Beasts of Holm were not then within official harbour limits (nor would they be until 1976). And Mr MacLennan argues, 'I cannot see how any scrap merchant could logistically, let alone surreptitiously, cut up a vessel of this size – salvage in its true sense implies a more or less intact recovery; raise it from the sea bed, even in sections, and then get it off the Island. Impossible . . .' Producing photographs to demonstrate 'how the power of the elements can remove and destroy any vestiges of a wreck,' John MacLennan concludes, 'Holm is a very exposed location, and the sea would soon pound a wreck to pieces.'

But there is real evidence that Grant was right. That the *Iolaire* is now a pile of demolished junk – 'you wouldn't know it was a ship,' commented one local diver, Jason King, in September 2008 – is scarcely surprising. Even before she foundered she was holed, torn, battered and collapsing. Moments before she sank, she split in two abaft of her funnel. Rear-Admiral Boyle reported extensive damage days later, from divers' descriptions and RNR study at low tide by the 'water-glass'. The site is extremely exposed and the *Iolaire* does lie in pretty shallow water, about thirty feet at low tide; the Admiralty chart indicates a depth of eight metres. Iron rusts – the nearer a wreck is to the surface, the more oxygenated the water and the faster rust will progress – and, even at sixty or eighty feet down, there will be considerable movement of wave, current and storm. The submerged wreck would fall apart faster than most imagine, and any human remains aboard – as was evident when the *Titanic* site was first filmed in the 1980s – would have dissolved completely, bones and all, and even in cold northern seawater, within two or three decades.

And a very experienced local diver, Chris Murray – celebrated winchman of the Stornoway-based HM Coastguard helicopter, who by his 2008 retiral had completed over 760 search-and-rescue missions – is adamant there is not nearly so much of the *Iolaire* on the seabed as there ought to be. 'She's right up against the rock – the plates of the bow-section, hard by the ledge. You wouldn't know it was a ship's bow, though. There's shells, ammunition, bits of crockery, stuff lying on the bottom. But the stern is definitely gone. I've read – or heard – somewhere – it was towed away; salvaged. A boiler is still there – it's a big boiler – but only one propeller, and the shaft cut.' I ask if there is any sign of the other propeller, the shafts, the engines. 'Nothing. None of that is down there at all.'

An amateur frogman might well retrieve mementos: a length of ship's rail, the odd button, an unbroken cup. But no skindiver could make off with a propeller, steam engines, shafts – all fashioned from valuable 'non-ferrous' metals, brass and copper and gun-metal, and precisely the most sought after for their scrap value. Mr Murray's findings are backed up by haunting video footage. Weighed alongside James Shaw Grant's recollections and the Admiralty's undoubted efforts to sell the wreck in January 1919 – the evidence in File 693 flatly confirms a done-and-dusted sale to Mr White; it was decided only to 'withhold operations' – there can be no doubt. Not long after the disaster, in an act of unspeakable desecration, the *Iolaire* was attacked for scrap.

According to Jason King, some locally believe that an abandoned boiler, still visible at low water from the Bràighe – the narrow isthmus of the Eye Peninsula – came from the *Iolaire*. A boat or barge, they whisper, could readily have hauled parts to this location, convenient for road access, and anyone seeing such operations there would not have associated it with the *Iolaire* at all. The actual retrieval could have been accomplished quite quickly. Any bones brought up – and they probably were – would have been chucked back into the sea.

But another senior Stornoway diver – who first descended to the wreck in the late 1960s – insists this boiler was not from the *Iolaire* at all, and argues that everything he saw on his early visits fits with the entire salvage of her engines – which may well have been sold for re-use, brass and gun-metal being pretty resistant to corrosion. He seals his testimony by producing extracts from Hansard – answers given to Dr Donald Murray in the House of Commons in 1919 and 1920 – confirming an authorised salvage operation just over a year after the disaster. (And he adds, though will not elaborate, that after some awful experience he swore he would never dive on the site again.)

J.J. MacLennan, consulting Pier and Harbour Commission records, is only aware of one other significant wreck in the area, the steam-drifter *Enzie*, which sank rapidly in October 1948 after a collision with a trawler. Chris Murray has found the nameplate of another craft, the *Nimble*, of which nothing can now be ascertained. And in all the years since 1 January 1919, there has been only one fatal grounding on the Beasts of Holm. On 25 February 1977, sailing in the middle of the night from Ullapool after a long and convivial session in the village pubs, the fishing-boat *Ivy Rose* struck and quickly sank, with the loss of two lives. A third man was rescued. At the subsequent Fatal Accident Inquiry, alcohol was heavily implicated.

Which brings us to the most abiding discussion of all: were the officers of the *Iolaire* drunk? That is not only the considered verdict in Lewis tradition – it was certainly the view of my grandparents and their contemporaries – but bears on the core of the tragedy. That the War Office and the Admiralty had made the most careless arrangements for getting these men home is an issue. That Walsh and Mason took a chance on ordering so many men aboard an under-crewed ship without nearly enough buoyancy aids or lifeboat space is undoubted. But none of this would have mattered if the *Iolaire* had made it into port, as both the *Sheila* and the *Spider* accomplished without incident in what, while a lively evening at sea, were by no means appalling conditions. The *Iolaire* disaster is one of grossly negligent navigation and the sobriety of her commander and his chief officer must be assessed.

It may be retorted that both Inquiries settled the matter beyond any doubt. The suspicion of drink, observed James Shaw Grant, arose not merely from the season of the year but 'probably was given an added impetus by the political background. Prohibition was a live and emotive issue at the time. Stornoway was on the run into its six dry years.' But, 'the Public Inquiry . . . disposed of the story pretty effectively. The jury's finding was not a simple negative, that they found no evidence of drink. It was the quite positive statement that no one on board was under the influence of intoxicating liquor. The jury included at least one of the leaders of the Prohibition movement in town who would not have subscribed to such a finding if he had any reason to suspect that the demon drink had a hand in the disaster.'

But the jury in February 1919 could only weigh the evidence actually heard in their presence and, besides, in a close community and amidst broken hearts and, perhaps, mindful of insinuations in the national press, may have been unduly anxious to declare that not even a passenger had tasted a drop of the hard stuff, the witching hour of New Year or otherwise.

But, despite their verdict and Grant's eagerness to vaunt it, there are some grounds for cynicism. First, the understanding of drunkenness in 1919 – both in society and in the Royal Navy – was rather different from our own in the twenty-first century, heavily conditioned by the drink-driving legislation of 1967 and general awareness of how even a small quantity of drink in the bloodstream can affect judgement, reflexes and spatial awareness. We all know that the legal blood-alcohol limit for driving, in the United Kingdom, is 80 milligrammes per 100 millilitres – a level many seriously argue is too high; and that the *safe* limit is no alcohol at all.

Before the advent of the breathalyser, the law reflected much cruder assessments from, really, the age of the horse – a doctor solemnly attended the police station and the accused was compelled, for instance, to walk along a straight line, stand on one leg, or recite such a phrase as 'The Leith police dismisseth us.' A man had to be the worse for drink – visibly swaying, disorientated, incoherent – and, in terms of Senior Service discipline, to be 'unfit for duty', which in the ranks at least meant merely unable to stand straight and do something with a rope. Boyle and Walsh both use a near-identical phrase for officers, in their telegram exchange of 5 January 1919 – 'fit state to take charge.'

But sailors must often have been on duty when they were easily over the modern lawful limit to drive a car, especially when we remember that it was only in 1970 that the Royal Navy abolished the daily rum-ration to all ratings below the rank of petty officer and aged twenty or older. This was a hefty dram – a gill of (admittedly watered) spirits – and had been abolished for officers in 1881 and warrant-officers only in 1918. Even sailors who did not indulge nevertheless drew it (at the behest, of course, of shipmates eager for more), and, no doubt, a few astutely stockpiled it. For that matter, a little rum was routinely issued to Great War soldiers minutes before battle.

Nor was there any prohibition on a naval officer drinking, in a moderate, responsible way, though some civilian shipping companies – such as the White Star Line – expressly forbade it on voyage. There was indubitably rum, at least, aboard the *Iolaire*; her crew would have had their daily issue of 'grog', and her officers would have had access to it, and perhaps other liquors, besides, for their mess.

Besides, whatever the fond conclusions of the jury at the Public Inquiry, we cannot absolutely exclude the possibility of temperate drinking among the passengers, at least, during her last voyage. There is no evidence that this was to excess, or that it was widespread, or indeed hard proof any man drank anything. In any event, they were off duty, going home for New Year, and had every right to enjoy a small refreshment if they chose. But it is naïve in the extreme to suggest it is impossible. Preparing his book in 1978, Tormod Calum Dòmhnallach gently pressed four of the few remaining survivors for their recollections.

Angus MacAulay, Shulishader, was emphatic. 'I didn't see anyone, as you'd say, smelling of drink.' Alasdair MacLeod, Coll, was no less firm. 'I didn't see any sign of drink on any man aboard.' Donald Murray, Tolsta, was more cautious. 'We didn't see anything that gave us that suspicion. For my own part, I didn't see the officers at all.' Donald MacRae, Ranish – who had been examined twice at the Court of Inquiry – was no less emphatic, pondering Dòmhnallach's question a half-century on. 'I didn't see anyone on the ship under the influence of drink. Anyway, I didn't see the officers when I went aboard, but none of us could have obtained drink when we were locked on the train. They couldn't have got drink, except what a fellow had in his pocket.'

There may certainly have been flasks in pockets, here and there, as Dòmhnallach concedes; and, as he also concedes, it would be natural enough, too, for men going home to enjoy a small libation at New Year. We do not know – and survivor recollections, in a strongly religious community, would undoubtedly be coloured by reluctance to recall anything that might allow the grieving to think young men had plunged into eternity with even a little liquor on their lips.

And in truth much evidence laid before the Inquiries, both as to the fitness of the officers and the course the Admiralty yacht sailed, is of very little value. Commander Mason and Lieutenant Cotter gave every impression of sobriety around seven p.m. Tuesday evening at Kyle, when Commander Walsh attested to their 'fit state to take charge' and they mingled cheerfully with a number of the Reservists. What, though, can that tell us of their condition five or six hours later, after the first magic moments of 1919 – as they ploughed their ship into a reef so notorious only one sad little fishing-boat has hit it in the nine decades since?

Likewise, Captain Cameron may well have endorsed the course laid before him at the Fatal Accident Inquiry – 'that's quite in order'; but he knew nothing, for sure, of the voyage the *Iolaire* actually took. From the moment she preceded him out of Kyle until he passed her stricken on the Beasts, Cameron never laid eyes on her, and the voyage plotted out for his scrutiny had been calculated by an Admiralty official from a few scraps of crew-testimony and a sneeze. The officers who did know were dead; their log, charts and records were lost in the sea.

And there are some real – if admittedly circumstantial – grounds for suspicion about Commander Mason; hard evidence the Admiralty, in January 1919, tacitly shared them. It seems odd that he should leave the bridge at one o'clock on Wednesday morning, with the *Iolaire* scarcely an hour from harbour. It is extraordinary that Cotter – who, in the last critical minutes, was alone, having sent James MacLean below and in touch with Leggett the helmsman only by speaking-tube – should not have sent for Mason. Calling the Master to the bridge is *de rigueur* even on a modern island ferry when a vessel is but half a mile from port.

As for Commander Mason's behaviour from the moment of grounding, it is incomprehensible. No orders are issued. There is not the least sign of any effort to take control, to impose order, to calm passengers and crew, to direct the lowering of boats or in any way to evacuate passengers, abandon ship and save lives.

It shall be borne in mind, besides, that a sailor under Royal Naval discipline who left his vessel wihout an express order to do so could be severly disciplined. Yet not one surviving passenger would attest in either Inquiry to seeing Commander Mason during those mounting minutes of emergency. Many could clearly remember Lieutenant Cotter firing rockets, lighting flares, shooting his flare-pistol. Mason does nothing, directs nothing, helps no one. He disappears. The evidence as to the entire disorder on board, the complete lack of direction, of men milling about not knowing what do, is overwhelming. And the man whose responsibility it is – to whom they all look for instruction, on whom their lives and their ship depend – is nowhere to be seen.

And only two crewmen – James MacLean and Leonard Welch – remember him at all. MacLean did not see Mason, 'but I heard him . . . He said, "What is that?" He was speaking to Mr Cotter.' Leonard Welch, trying frantically to coax his wireless equipment into life, did lay eyes on his captain. 'The Commander came down to the cabin and told me to send a distress signal. I told him the wireless was smashed and I tried to rig up an emergency set.'

This is important, for Leonard Welch was one of the Court of Inquiry witnesses 'who saw, or spoke with, any of the officers' duly recalled, at Rear-Admiral Boyle's orders, for further questioning as to their condition. The rumours were notorious. The insinuation was evident. The questions to be put seem obvious. But Leonard Welch's second interrogation is remarkable.

1126. Q. Did you see the Commander during the passage?
A. I saw him shortly after she struck.

1127. Q. Did he speak to you?
A. Yes, he told me to send out a distress signal.

1128. Q. Was that the only time you saw him?
A. Yes, that was the only time.

1129. Q. Was there anything unusual in his manner?
A. The ship had struck then and she was breaking up. He seemed agitated.

1130. Q. Was it the consequence of the condition of the vessel or otherwise?
A. The consequence of the condition of the vessel, I should say. Everything was collapsing, all round.

1131. Q. You did not see him alive again?
A. No, not again.

The shaken telegraphist of 1919 quite contradicts the commercial traveller of 1942: Welch attested to the Court of Inquiry that Mason was not 'calm' but 'agitated'. Yet this was never a line of questioning likely to nail down one elementary point: had Commander Mason been drinking? And what should really make us suspicious is Leonard Welch's absence from the subsequent Fatal Accident Inquiry. He was not called. His evidence was not heard. Neither Crown Counsel, Mr Fenton, nor the solicitor for some survivors and families, Mr Anderson, had opportunity to question the one man who saw Mason after his ship ran aground, *the one man close enough to smell his breath*. It is possible – the Spanish 'flu raged – that Welch was ill. It was possible he had been sent back to sea, somewhere far away in His Majesty's service. And it is possible – a matter of removing a few pages from the copy of File 693 put at the disposal of the court – that none of the lawyers was aware of his evidence or even that Leonard Welch existed, a man who might have found Anderson's questions, at least, much more searching than those of the Court of Inquiry.

At this point, too, Rear-Admiral Boyle's contemptuous reference to the first informal questioning of all survivors 'at the Base on Tuesday 7th January, with a view to weeding out those whose evidence was of no value' could assume another significance. It might certainly be taken as reference to those who had seen nothing of importance and had no evidence of value to give. But it would also identify those whose testimony might mortify, granting time enough to discard some witnesses or ensure that any line of questioning pursued on the record – and Boyle was as consciously establishing that paper-trail as pursuing an investigation – was phrased as carefully as possible.

There have long been claims of an *Iolaire* cover-up; of some tricks at the Court of Inquiry and, perhaps, a degree of Establishment collusion in later public proceedings. The inexplicable absence of Leonard Welch from the Fatal Accident Inquiry does not, of itself, prove anything. But it is much more significant than recent claims of tape-recorded, gossipy conversations with old men long dead. And what we have noted does strongly suggest that, by 01:55 on New Year's morning, Commander Mason was in no fit state to take charge – he notably failed to do so; we have to admit alcohol as a very strong possibility for that condition (while acknowledging shock, terror and exhaustion might have been others); and we have some reason to believe the Naval authorities had real grounds to fear revelations in open court.

We may now diffidently trace the last voyage of HMY *Iolaire*. The precise course is inevitably a matter of conjecture, and can never now be known. We know only where it ended. We do have three clear theories. And there are two important caveats. There is a little evidence that on occasion the Admiralty yachts did make unorthodox passage by Kyle to Stornoway, coming deliberately by the Point coast and the Bràighe on an approach to Arnish not very different from that of the present Ullapool ferry. Captain John Smith's father remembered this happening on one or two occasions during his own Great War leaves. It may have been judged an easier course to navigate, at the cost of some additional time and fuel; there may have, besides, been a tactical wariness of U-boats, lurking perhaps in the sea-lochs of eastern Lewis. But it would be a distinctly dangerous passage in the dark, with the hazard of Chicken Rock as well as the Beasts of Holm to complicate matters.

It is also not widely appreciated that a ship's compass is affected by her own iron and steel. So every vessel has a different 'compass course'; her particular eccentricities established at trials and the compass-card set to allow for them. And, to compound our speculation, magnetic North is not a constant, and has, indeed, shifted since 1919.

On 23 January 1919, the Admiralty Director of Navigation – J.A.

Welsh – examined the scant evidence and recorded his conclusions for Rear-Admiral Boyle: the *Iolaire*, having set a course north from 12:30 to the centre of the Stornoway Harbour entrance, was in fact steered in the twelve-mile run thereafter to about six cables eastward of it; and that, realising off the coast of Lewis he was significantly to the east of his intended position, the 'Commanding Officer' altered course to come in close by Arnish Beacon Light. This led him dangerously close to Holm Point, the drizzling rain having perhaps affected the brightness of Arnish Point Light and leading him to believe the *Iolaire* was nearer that lighthouse than she actually was.

Half a century later, Captain John Smith, a native of Ness and for many years master of the *Loch Seaforth* and her first car-carrying successors, made a detailed analysis for BBC Gaelic radio. His theory, to my mind, is the most convincing and fits most readily with the known facts and the evidence of most *Iolaire* survivors – that the wrong course was set at 21:55 on the evening of 31 December, when off the north point of South Rona. The *Iolaire* thus sailed too far to the north-east and then – the situation compounded, perhaps, by a new officer going on watch at 01:00 on New Year's morning – some floundering over a lighthouse. It is additionally possible, as Alexander Reid has suggested in evidence for the Smith hypothesis, that Mason or Cotter or both actually mixed up the relevant lighthouses.

But Reid – architect with the new Western Isles Council and only an amateur yachtsman, and in 1978 studying the question for an appendix to Dòmhnallach's Gaelic essay on the tragedy – opted for a third theory: his own. He argued that things went wrong when the *Iolaire* resolved to overtake the fishing-boat *Spider* – the manoeuvre, perhaps, of a tired man in a hurry, but a frankly irresponsible one in the middle of the night, in fresh to gale-force conditions, by an officer so entirely lacking in local knowledge. And Lieutenant Cotter would struggle, besides, in the critical minutes of James MacLean's absence from the bridge and, of course, without his Commander. Thus 'a straightforward navigational exercise' became 'a complex problem, involving the mental calculation of relative courses, speeds, and position, highly taxing in that night's conditions . . . if *Iolaire* had reduced speed to five knots when half a mile behind *Spider* and followed her in, she may perhaps have been delayed by fifteen minutes.'

Reid does not care for the Admiralty explanation – 'oversimple and an inadequate explanation for that degree of incompetence in a Naval officer', and asserts that Captain Smith's explanation 'would require an incompetence or incapacity on the part of all officers and crew in the wheelhouse for the last hour and a half which is difficult to accept.' But the course Alexander Reid proposes ignores considerable eyewitness evidence – not least by passengers from Point, who knew their own shore, and besides by James MacLean of the vessel's crew – of land long close to the port side, supporting surely that the *Iolaire* rushed southwards to Holm Point by a run from Bayble. Nor is it at all clear, from 1919 testimony, that the *Iolaire* 'overtook' the *Spider* – that is, that she consciously overhauled her from behind: certainly, she did not fall ahead into the same course for harbour. Neither Donald Murray or James MacLean mentions a fishing-boat at all and the Court of Inquiry statement from the *Spider* skipper, James MacDonald – worth again quoting in full – is by no means emphatic.

> During the latter end of December, 1918, I came to Stornoway on the fishing boat *Spider* to prosecute the herring fishing. On Tuesday, 31st December last, we proceeded to the fishing grounds in the Minch and shot our nets at the Shaint [sic] bank. About 10 p.m. we hauled our nets and made for Stornoway. When sailing past the mouth of Loch Grimshader a Steamer passed us on the starboard side, apparently on its way to Stornoway. I did not identify the ship and suspected it to be the *Sheila*. We followed immediately in the wake of the Steamer, and when approaching Arnish Light House I noticed that the vessel did not alter her course when passing the light, but kept straight on in the direction of the 'Beasts' at Holm. I remarked to one of the crew that the vessel would not clear the headland at Holm as it went too far off its course to make the Harbour in safety. Immediately afterwards we heard loud shouting and then knew that the vessel was on the rocks. We were passing the Beacon Light at Arnish at that time, and could hear the shouting of the men as we were coming into the Harbour. The night was very dark and a strong breeze from the South was raging and a heavy sea running. We

were unable to give any assistance as we could not rely on our engine to operate in such a rough sea.

On our arrival at Stornoway we reported to the Naval Authorities that a Steamer was on the rocks at Holm.

Truth.

MacDonald's description of the encounter is of the momentary convergence of two ships seeking the same port, but on different courses. The *Iolaire* 'passed us on the starboard side'; there is no suggestion that she overhauled the little smack from behind, and he explicitly says the yacht then proceeded on a different bearing to his own. This is not an overtaking manoeuvre and it is quite possible neither Cotter nor Leggett even noticed the fishing-boat.

There is little here, then, to found Reid's hypothesis, and his seeming confidence in the basic seamanship of Mason and Cotter is most odd: the wreck of the *Iolaire* by definition attests to gross 'incompetence or incapacity', so absurd in its details and so awful in its consequences that Lewis has struggled for almost a century to process it.

No mariner is better placed to comment than Captain Angus MacKenzie, who served for several years as a deck-officer with Captain Smith on the *Loch Seaforth*, as master of her car-carrying successor and subsequently, from 1978 to 2004, as Harbour Master of Stornoway – intimately acquainted with the passage from Kyle and, in all their complexity, the approaches to the harbour.

'I wouldn't make anything of the collision with Kyle pier,' says MacKenzie. 'Many a time John Smith banged it; it's not significant.'

And the way they described courses and bearings then isn't what we were used to. Ships have gyro-compasses now; we talk in degrees. Of course Smith was used to the magnetic compass, quarter-points and so on. But he was absolutely right – the course this fellow MacLean describes is completely wrong. It would take them straight to Tiumpan Head. And Mason's behaviour makes no sense at all. We had a point marked on the charts – an island at the mouth of Loch Erisort – and that was when Captain Smith was to be called to the bridge, when we passed that. And the carry-on after she grounded, the Commander's behaviour – well, there was something far wrong with that man, whatever it was.

He was RNR, of course – a reservist; not Royal Navy, not Merchant Navy. There was a lot of that even in the last war – someone middle-class, with a wee bit of education, and who had maybe a yacht, and they would make him an officer, but they weren't really qualified at all . . .

We can hypothesise.

It is the last day of 1918, and Rear-Admiral Boyle is not happy: he knows many ratings must be collected from Kyle of Lochalsh, but none of his colleagues in authority can tell him precisely how many or when they will arrive. In a brisk exchange of telegrams with Kyle, Commander Walsh and he agree that the *Sheila* is unlikely to cope and Boyle resolves to send the *Iolaire*, a small, slow old thing that has only been at Stornoway since October. Half her crew – largely English – are still on Christmas leave. The available deck-officers, Richard Mason and Leonard Cotter, have shipped into Stornoway, but not many times, and seldom by night, and neither has ever been the Officer of the Watch on a passage into Stornoway in darkness. Perhaps aware of this inexperience, Boyle has expressly instructed they are not to berth by the Stornoway quayside; they will drop anchor in the harbour and HMD *Budding Rose* will flit the libertymen ashore.

HMY *Iolaire* sails around 09:30. She reaches Kyle of Lochalsh at 16:00. Berthing is messy – she thumps the pier hard. Were vast ranks of libertymen already waiting, she could load and turn round within half an hour. It would still be a voyage in darkness. She might still go aground. But with a southerly wind still very light and with Stornoway, around 22:30 or 23:00, merry rather than unconscious, many more would have survived. But the trains have not come to Kyle, and fussy Commander Walsh wants every last available berth on the *Sheila* taken up. Captain Cameron cooperates, only mildly impatient. Commander Walsh wants to know if the *Iolaire* could, if necessary, take 300 men. She can easily do so, says Mason; they have a 'short discussion' about boats, lifebelts, and the weather. They both know they are bending the rules.

Everyone hangs about. At 18:15 the first part of the train arrives; the ratings – 'slightly disorderly from a point of view of discipline, but otherwise they were all right' – are efficiently paraded, Lewismen separated from Harris and Skye comrades. Ninety-five files for Stornoway are sent aboard the *Iolaire* by her single narrow gangway. A hundred and ninety, so far. The Harris ratings are ordered to the Red Cross Rest; no boat will collect them until Thursday morning. Eleven, though, somehow get aboard the *Iolaire*, and seven of them will die. Some Lewismen, too, jump ship from the *Sheila* to the *Iolaire*; one or two make the opposite transfer.

The rest of the train arrives at 19:00. This time there are 130 men for Stornoway, and Walsh has just marshalled them when he learns the *Sheila* can take sixty more libertymen. Thirty files are immediately sent to the mailboat; the other seventy men are hastened onto the *Iolaire*. Her master is in jolly form. He greets men, tells them to make themselves comfortable, invites a few to stow their kit in the chart-room, where it will be dry. For most survivors, this is the only time they will lay eyes on Richard Mason.

Walsh at last learns the entire train has come and he need expect no more passengers. He goes along the pier, watches the last libertymen board, chats to Commander Mason – who says the glass is rising and he expects to make a good passage. His command can do about ten knots. The officers bid one another a Happy New Year and goodnight. A few minutes later, about 19:30, the *Iolaire* sails. It is darkest night, clear, a southerly breeze astern, but things will liven up. Mason and Cotter are dangerously short of manpower, with half their hands away. There should be at least one fixed lookout for'ard, as Commander Bradley confirms at the Public Inquiry, stationed at the yacht's bows or in the crow's nest – if there be one; all the more important when the vessel's bridge is so far astern. But James MacLean will attest vehemently he was 'the only one keeping a lookout. There was no one stationed at the boat's head, not during this voyage at all . . .'

At 21:55, the first mistake is made. The *Iolaire* is sailing past the northern point of South Rona, when course is to be altered to make a straight bearing to Stornoway – 'course was then shaped North 2 degrees East,' Commander J.A. Welsh, Director of Navigation, will assert in his 23 January analysis, drawing from James MacLean's testimony: the Campbeltown seaman finds her on that course when he goes on duty, and of all involved in the ship's navigation only he will live. But, Captain John Smith will point out many years later, this course from South Rona was 'wrong in the first place . . . the correct line from the north point of South Rona to Arnish Light is in fact North, seven degrees West. If she held course from 21:55 hours to 00:30 hours, this would place her in a position eight and a half miles South, 70 degrees west from Kebbock Light . . .'

HMY *Iolaire* is sailing just a little east, but enough to draw her steadily off her proper heading – making, in fact, not for the centre of Stornoway Harbour, but towards Tiumpan Head, the lighthouse at the end of the Eye Peninsula, the most easterly point of Lewis.

James MacLean is roused from his berth about twenty minutes before midnight. He has had less than two hours' rest. He takes the wheel at 00:00, and will steer the *Iolaire* for the next hour. He cannot see Commander Mason, who is standing on the bridge immediately above, but they are in contact by speaking-tube. MacLean can hear his orders clearly. When MacLean assumes the helm, as he will tell the Court of Inquiry, the ship is 'steering north-easterly – just a touch easterly.' (Alexander Reid will assert in 1978 that this must be a transcription error, that the hyphen marks a hesitation rather than a precise compass point – 'steering north . . . easterly, just a touch easterly,' but this seems a faintly desperate argument.) MacLean is ordered to alter course at 00:30 – 'to North'. That bearing is held until he is relieved half an hour later, at 01:00. He can see the lights of Arnish Point from his vantage, bearing 'about half a point on the port bow . . . about 12:30. It is 'about the same place' when Ernest Leggett takes over. MacLean will also give evidence that at no point were bearings taken while he was on duty at the wheel; but he cannot see what Commander Mason is doing above, and he has been explicitly instructed to alter her heading.

She is now five miles off course and, in ordering a 2-degree correction to port at 12:30, Commander Mason has made an elementary but deadly mistake. He naturally wants to centre the *Iolaire* on Arnish Light. He has, probably, drawn quick bearings, now confident – with Milaid Light, Kebbock, flashing well to port – that he is well clear of the rugged east coast of Lewis, and can safely head straight to Stornoway. *But Mason has mixed up the lighthouses.* Miles to the east now of her intended bearing, the *Iolaire* is at the limits of Milaid's visible range. Mason has not even noticed the dimmest flicker far to port. The light he thinks is Milaid is Arnish Light. The light to which he is conning the *Iolaire* – he 'knows' it is Arnish – is, in fact, the lighthouse on Tiumpan Head. All three lighthouses have a different flashing sequence – Arnish once every ten seconds, Tiumpan two flashes every thirty, Milaid one flash every fifteen – but Mason is not a local officer, he is relaxed and confident, does not check his charts, and – well, it is New Year – has perhaps indulged in a drink.

MacLean will later say that Commander Mason was still on watch when he left the wheelhouse, but the seaman loiters on the promenade deck, and Lieutenant Leonard Cotter assumes the watch. MacLean joins him on the bridge minutes later. At a critical moment in the voyage, the Captain has left the bridge and gone below; a new helmsman and a new officer are in charge. And the weather is deteriorating: the wind is rising, there are frequent showers of drizzling rain, diffusing (and thus brightening) the beam of assorted lighthouses, the only fixed points in the blackness engulfing them as the *Iolaire* washes on.

Lieutenant Cotter has taken the watch sharing all Mason's assumptions. He is, besides, tired. It is a long day. He is also short of hands, and has only James MacLean to assist on the bridge. Leonard Cotter may have grounds quietly to worry about the present fitness of his commanding officer. Minutes later, he has still more – Arnish Light flickers and disappears. Cotter can hardly believe his eyes. But it is – we know – the light at Tiumpan, and as the *Iolaire* chugs blithely towards Bayble the rising landmass of the Eye Peninsula has loomed invisibly in Cotter's sightline from the bridge, 'closing' the light of Tiumpan Head. On deck, Donald Murray – who has quite correctly identified the true Arnish Light and the entrance to Stornoway Harbour – cannot comprehend why the *Iolaire* has sailed past it. (And he has seen no fishing boat either; in fact, there is no witness evidence on board that the yacht 'overtook' the *Spider*.)

Cotter may besides now lose the light at Arnish, the real one, 'closed' by Holm Point. He is disturbed, preoccupied. MacLean has joined him. To minimise embarrassment, Cotter gets rid of him. As MacLean will relate, the seaman is sent below to rally the hands for the anchor; harbour, he is told, is not far away. Leonard Cotter should call Commander Mason, but does not do so – embarrassment? fear of challenging earlier navigation? unease as to his state? Instead, he quickly instructs a marked alteration of course, just after 01:25, well to port – to west – but gentle enough not to be felt by MacLean or anyone else, nor in such darkness to be seen. James MacLean is not on the bridge to hear a new course ordered, nor in the wheelhouse when Ernest Leggett receives it. Lieutenant Cotter now maintains his course for Arnish Light, seven and a half miles ahead. He passes Chicken Head, crosses the bay by the Bràighe. He still does not call Commander Mason.

On the starboard deck, many of the libertymen grow aware of land – the smell of it, the boil of surf, the merest flicker of white waves breaking. Some grow uneasy. They know this is not the orthodox route. James Montgomery will say he saw land five minutes before impact – the point east of the bay that would be the grave of the *Iolaire*. John MacKinnon of Tarbert saw land for the preceding quarter-hour. John MacKenzie of Portvoller thought, watching from the rail, she was sailing too far to the leeward. James MacLean notices land, too – twenty-five minutes before impact. Kenneth MacLeod of Swordale and his mates, seeing rocks awash so close, exclaim, 'He's running aground . . .' Several feel another change of course, the *Iolaire* rolling markedly.

Norman MacIver of Arnol, who will escape by James MacLean's direction, will be adamant at the Fatal Accirent Inquiry. 'When I was on the starboard side of the ship I could see land for quarter of an hour before she struck. She was going along the land all the time. I reckon that land was five hundred to six hundred yards away when I saw it on our starboard side. At that time I was not taking any notice of the lights as I did not know them. There was a beacon light just ahead of here when she struck the Rock.'

Cotter has two immediate navigational problems: one perhaps the distraction of the *Spider* – it is by no means clear whether he overtakes her, dodges her, falls away from her, but he may be aware of her and is eager to keep his distance – and his obsession with Arnish Light. He is navigating by eye and compass with no spare hand – MacLean is below again, readying the men for the anchor. Cotter is anxious to keep Arnish Light open, trying to get the town-lights of Stornoway, possibly keeping well to starboard of the little *Spider*. Holm Point, seventy-five feet high, has materialised before him, blocking Arnish Light, He steers round it, very close, too close. He eases in towards the shore. He can see a glow from town. Arnish Light is 'closing' again. He orders a last change of course to port, seconds before all his assumptions and his world come to an irrefutable, crashing end.

The rest we know. Cotter is shocked, near-stupefied, not sure what to do, not at all certain where he is, but surely near town, from whence help must quickly come. His immediate instinct is to ring down 'Stop'. The engines are indeed stopped and, with rapid flooding and expiring boiler pressure, the firemen flee. Within moments one option – pulling astern, hauling the *Iolaire* off, trying perhaps to beach her on an available shingle cove for readier evacuation – has been blithely forfeited. No anchors are put out; no effort is made to stabilise the heaving, bouncing ship. An aghast Leonard Welch finds his wireless equipment is useless.

Commander Mason materialises briefly. He seems . . . out of it. No orders are given. No order is established. Cotter does things because passengers tell him to – fires rockets, burns flares, signals with a lamp no one on land can see, blows the whistle. It does not occur to him to fire his ship's three most noisy guns. It happens that the first, near-silent rocket is blue. So is the second. On watch at the Battery, William Saunders reports – quite correctly, per training – that a vessel requires a pilot. Time is lost. When he reports a red light, the pantomime ashore begins; but it is academic: snug in the Imperial Hotel, Boyle has made no effort to get men immediately to the scene and to assess the situation. By the time the alarm is raised, it is probably too late. Lieutenant Wenlock and the *Budding Rose* wade nearby as Cotter fires the last red flare, as HMY *Iolaire* disintegrates into the sea. Saunders, alone, dogged, remains on watch from Stornoway. 'Just after three o'clock,' he will tell the Public Inquiry, 'the masthead light disappeared.'

The *Iolaire* disaster may well put a new iron into the Lewis soul; inspire young men to press for change, resist manipulation and dead traditionalism. But hundreds will simply leave. Lewis will never again look at authority and the legitimacy of government in quite the same way. Her outstanding part in the Great War has been mocked, tainted, defiled. In the most immediate and personal way, dozens of homes are bereaved in the cruellest fashion: dozens are widowed, hundreds are orphaned. The island is devastated. Ninety years on, few Lewismen can yet utter the name of the *Iolaire* without emotion.

The survivors find themselves marked men, whom many – especially those who have lost a loved one – can never quite look at in the same way again. Like other Great War veterans, and more so, they find themselves distanced from those who have not served, distanced from their parents, distanced from their own children. Those who came living to shore will gain not a penny from the Iolaire Disaster Fund. A number will emigrate. Some, at least, will leave for the mainland, not to return. But, wherever they are and whatever they do, they plug on with their lives, diligent and useful, if a little quieter than most, and several unhesitatingly serve again in Hitler's war.

In the late decades of the twentieth century they will find their voice, an articulate serenity; realise that people want at last to hear their story. They come together, in little groups, on radio. There is a tenderness of fellowship. There is due solemnity, a disciplined anger, a honed, patient acceptance. But there is that bond between them, that thankfulness, and a certain very quiet pride. They are the rugged, rather splendid Edwardians who, by grace and courage and resilience and the weight of years, become legends. Others, too, had won the Great War. But they are the men who survived the *Iolaire*.

One, Donald Murray, veteran of Antwerp and the Dardanelles and the bobbing boats of the *Lusitania*, who sailed on the *Iolaire* and yet lived, will take my hand in August 1988 and say, 'I went through great experiences . . .'

Another, in the summer of 1965, comes to see the memorial at Holm. John Finlay MacLeod has not once been here since he stumbled through the night to Stoneyfield Farm. He sits now on the sward with John Murray, *Iain Help*, the first man ashore by his lifeline, and John Finlay's son, John Murdo, takes a photograph of those Sabbath-suited, thoughtful old men, gazing out over the Beasts of Holm. They have their fill, their memories, the recollections of lads long drowned, and at length leave. John Finlay will never go there again. By next summer, John Murray is dead.

Though he got, in 1921, his medal and parchment from the Royal Humane Society, MacLeod will never be honoured by the Crown. He will die on 21 December 1978. Yet, on the brink of a new century, his granddaughter shall be an appointed dressmaker to The Queen, who will wear a Sandra Murray creation when, on 1 July 1999, she opens a renewed Scottish Parliament.

In 1987 a shrewd, spry Free Church elder in Lionel, Ness – Norman Smith, himself a hero of the trenches of Anzio – is involved in the production of a new 'Roll of Honour', an illustrated volume listing all from western Lewis who served in the Second World War. And he approaches two quiet elderly crofters, last of their company this side of the island, to supply a little postscript.

> Although our names do not merit an entry in this record we, the remaining survivors of the *Iolaire* tragedy of the 1914–18 First World War, are pleased to see the publication of this record for the 1939–1945 Second World War for the Isle of Lewis, from the Butt of Lewis to Bernera, Uig, and we feel honoured to be asked to add these few words to the publication.
>
> Many events of bygone days are sacred in the memory of those who took part in them and we are sure that this book will bring back such memories to many whose names are included in it.
>
> We would like to thank all who took on the onerous task of putting this record together and we hope that through its success they will feel rewarded for their work.
>
> DONALD MORRISON
> 7 Knockaird, Port of Ness
>
> JOHN MACLENNAN
> 4 Aird Uig

When Donald Morrison, Am Patch, dies in July 1990, there will pass the last who can discuss or clearly recall that last voyage from Kyle.

Of all who made it off the ship alive, only his gravestone will bear an inscription of honour, 'Survivor of the *Iolaire* Disaster.' Speaking on my own modest radio programme in January 1989, he will say, 'I can't forget it. I think about it every day of the week. We were only a few minutes from harbour – you can't forget a thing like that . . .'

Regularly Norman Smith collects the Great War veteran of a Sabbath evening for the Lionel prayer meeting, a weekly date the Patch will keep till the last months of his life. Now and then, the concluding stanzas of Psalm 37 are now and again sung in church:

He in the time of their distress
their stay and strength doth prove . . .

And Donald Morrison will always say, as Norman drives him home, 'You know, whenever I hear that verse, I'm back on the mast – and I didn't even know it was in the Bible . . .'

And in the secret places of his heart he would marvel at Providence – the grace, the awful mystery, that spared him and took his brother, took so many; that purpose that at the last minute placed my own great-grandfather on the *Sheila* – that might besides, or instead, and by mere impulse of hand on a ship's wheel, have seen the *Iolaire* rejoicing into Stornoway.

John MacLeod
Marybank
Isle of Lewis
31 October 2008

Banntrach Cogaidh

Sgaoil i aodach air gàrradh:
briogais bhàn is lèine gheal,
crios gorm leathann,
còrd geal caol,
Bha mi air clach ri taobh,
bonaid cruinn air mo cheann:
bonaid m' athar.
Deise ghorm na stiallan
mar reub iad bho chorp i
nuair fhuair iad e
fuar bàthte air an tràigh,
air a pasgadh le làmhan gràidh
is cridhe brist':
mar deise rìgh dol gu banais.

Carson tha thu dol uair sa mhìos
don phost-oifis nad aodach dubh,
le cridhe trom,
's a' tilleadh feasgar
le leabhar a' pheinnsein nad làimh?
Carson a thog thu mi nad uchd
's do cheann crom,
is fhliuch thu m' aodann le do dheòir
nuair thuirt mi,
"Mhàthair, cà'il m' athair?"

Aon là san sgoil
sheas sinn sàmhach dà mhionaid
a' cuimhneachadh
air laoich a' Chogaidh Mhòir,
is ruith mi dhachaigh na mo dheann
a dh'innse dhi
"Cha bhi cogadh tuilleadh ann."

Oidhche gheamhraidh bha i snìomh;
shuidh mi ri taobh.

War Widow

She spread his clothes on a dyke:
light-coloured trousers and white shirt,
a broad blue belt,
a thin white cord.
I was on a stone by her side,
a round bonnet on my head:
my father's bonnet.
Blue uniform in tatters
as they tore it from his body
when they found him
cold drowned on the beach,
folded with loving hands
and broken heart:
like the clothes of a king going to a
    wedding.

Why do you go once a month
to the post office in your black clothes,
with heavy heart,
and come back in the evening
with pension book in your hands?
Why did you bury me in your bosom
with your head bent
and wet my face with your tears
when I said,
"Mother, where's my father?"

One day in school
we stood silently for two minutes
remembering
the heroes of the Great War,
and I ran home pell-mell
to tell her
"There will be war no more."

One winter's night she was spinning;
I sat down by her side.

Bha ceann dol liath 's i fhathast òg.

An lùib an t-snàth chaidh fuiltean
    mìn
mar shìoda measg an duibh.
Thuirt i, "Glèidh cuimhn' orms'
nuair chì thu m' fhalt an lùib
    an t-snàith
's nach bi mi ann."
'S mar thubhairt, bhà.

Rinn iad d' uaigh ri taobh nan tonn.
Cha chlisg thu chaoidh aig gaoith
    no stoirm.
Sibh sin cho rèidh, thu fhèin 's
    an cuan –
cha toir e tuilleadh uat do ghràdh.
Nach math gun tug am bàs thu
    tràth
's nach fhac' thu cogadh ùr nad là
's nach fhac' thu mise falbh don
    bhlàr
le deise ghorm is bonaid cruinn
mar bh' air m' athair
nuair fhuair iad marbh e
aig a' Bhràigh.

An t-Urr Iain MacLeòid
(1918–95)

Her hair was going grey when she was still
    young.
Amidst the thread went one fine hair

like silk among the black.
She said, "Remember me
when you see my hair amidst the thread

and I'm not there."
And what she said came true.

They made your grave beside the waves.
You'll never flinch at wind or storm.

You've made your peace there, you and the
    sea –
no more will it take from you your love.
It's good that death took you quite soon,

that you saw no new war in your day
and did not see me leave for battle

with round hat and blue uniform
like those my father was wearing
when they found him dead
at the Bràighe.

Rev. John MacLeod (1918–95)
translated by Ronald Black

Appendix I

CHARTS

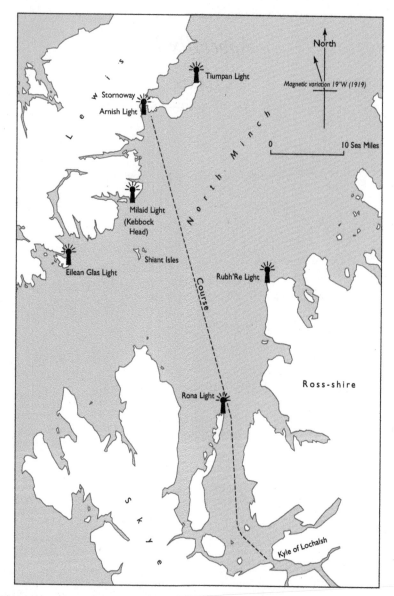

The correct course from Kyle of Lochalsh and the Inverness railhead to Stornoway, by which the Lewis port was served (with a southern dog-leg to Mallaig and trains for Fort William and Glasgow) till March 1973, when the present 'short-sea crossing' was introduced to Ullapool by the 1970 car ferry *Iona*. The major lighthouses are shown. (After Alexander Reid, 1978)

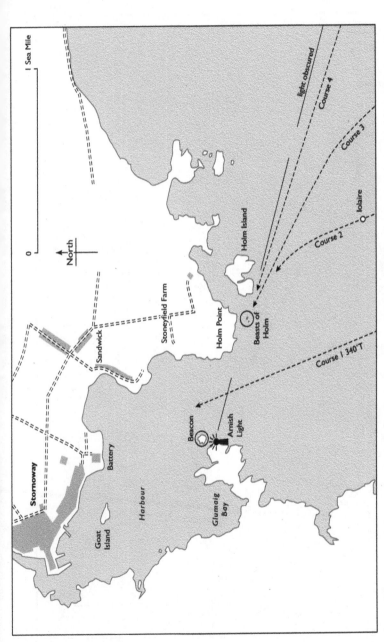

Approaches to Stornoway, 1919, showing lights, hazards and points of reference in the tragedy such as the Royal Naval Reserve Battery and the survivors' first refuge at Stoneyfield Farm. The difficulties of bringing the heavy, horse-drawn life-saving apparatus quickly to Holm Point are evident, as well as the entire exposure of the locale to the southerly gale.

Course 1 – The proper passage into Stornoway from Kyle. Course 2, postulated: Alexander Reid's theory – the *Iolaire* sails calamitously off-passage as she overtakes the little fishing-boat *Spider*. Course 3 – the Admiralty theory: how the *Iolaire* went off course in the Minch. Course 4 (and the most credible explanation) the hypothesis of Captain John Smith – the *Iolaire* on a correcting westward run, too close inshore, after setting the wrong passage from South Rona light. (After Alexander Reid, 1978)

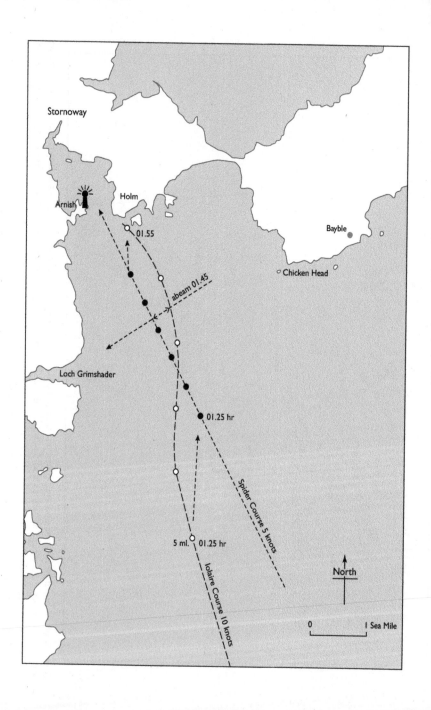

Stornoway

Arnish

Holm

Bayble

Chicken Head

01.55

abeam 01.45

Loch Grimshader

01.25 hr

Spider Course 5 knots

5 ml.   01.25 hr

North

Iolaire Course 10 knots

0   1 Sea Mile

The *Spider* Theory (left)

In a hurry to be home, the *Iolaire* overtakes the much slower fishing smack *Spider*: alone on the bridge, navigating by sight rather than compass, unfamiliar with tide and currents and feeling his way to Stornoway in the dark, Lieutenant Leonard Cotter fatally miscalculates and sails the *Iolaire* onto the Beasts of Holm.

There are three difficulties with this hypothesis. There is no plain evidence that the *Iolaire* even saw the *Spider*, far less deliberately overhauled her. It is counterintuitive that Cotter would have proceeded on a course so at variance with that of a vessel he had just overtaken and obviously proceeding into the same harbour. And there is considerable survivor testimony that the *Iolaire* sailed for many minutes close to the land of Bayble, Chicken Head and the promontories and islets of Holm – in other words, a run almost due west from Bayble and a very different passage to that outlined here.

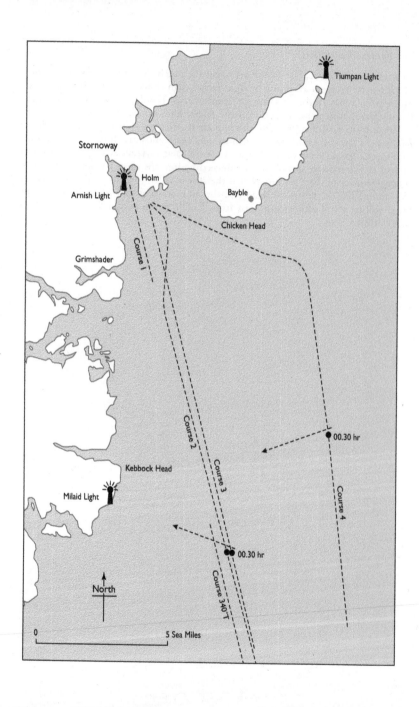

Tiumpan Light

Stornoway

Holm

Arnish Light

Bayble

Chicken Head

Grimshader

Course 1

Course 2

Course 3

Kebbock Head

Milaid Light

00.30 hr

Course 4

00.30 hr

North

0   5 Sea Miles

Course 340 T

The Smith Theory (left)

Course 4, plotted here, accounts most credibly for the disaster – as proposed by Captain John Smith (1908–1985), a native of Ness in northern Lewis and for nineteen years the master of MacBrayne's *Loch Seaforth*, which served Stornoway from Mallaig and Kyle between 1947 and 1972. He was the first Lewisman to skipper the route and, under his command, she never once missed a crossing due to weather conditions. He also captained her immediate car-carrying successors, the *Iona* and the *Clansman* – latterly on the new route to Ullapool – until his retirement late in 1973. (Shown here besides are the correct passage into Stornoway – course 1; Reid's *Spider* theory – course 2; and the route to tragedy suggested by the Admiralty's Director of Navigation, course 3.)

Having set the wrong course from the South Rona light (course 4), the *Iolaire* proceeds north and further and further to the east of the correct route, perhaps compounding the error by mistaking the light at Tiumpan Head for that at Arnish. Once the mistake is apprehended – probably when the Tiumpan light is 'closed' by the looming landmass of the Point peninsula – the *Iolaire* alters course decisively to port, running westwards from off the shores of Bayble. Now obsessed with the Arnish light, Lieutenant Cotter pays scant regard to hazards on his starboard bow. At 01:55 hrs, far too close to shore, alone on the bridge and with no forward lookout, he smashes his command into the Beasts of Holm.

There is strong supporting evidence for Captain Smith's theory – the testimony of James MacLean, the only surviving helmsman; the witness of a number of surviving passengers as to the land and breaking waves seen to starboard well before impact; and plain chronology: by this course alone, the *Iolaire* goes aground at just the properly calculated time from her Kyle departure. (After Alexander Reid, 1978)

Appendix II

HMY *IOLAIRE*
1 JANUARY 1919

Lloyd's Register of Yachts, 1913

Name: *Amalthaea* ex *Iolanthe* ex *Mione* ex *Iolanthe*
Material of Build: Iron
4 B.H. Cem.
Type: SchSc (schooner rig screw-powered)
Lapthorn

Tonnage: (i) 204.08 (ii) 414.86 (iii) 634

Dimensions: 189.3' (length) x 27.0' (beam) x 15.0' (draught)
Stern Length 6 ft.91
Eletric Light ND (no date) 1902

Builders: Ramage and Ferguson

Place: Leith

Date: 1881
4 months.

Machinery: C.1, 2 Cy. 27" & 50" – 30"
108 NIP 1B 3pf GS55
HS1804. 80 lb. DB. 75 lb.
M Fauld & Co., Dumbarton

Owners: Sir Charles G Assheton-Smith, Bart
ss Son. 2nd No. 3 – 9,06
ss Bng. No. 1 – 11

Port of Registry: London.
N.B 11,98

Mercantile Register

GR5

ON 85043

Name: *Iolanthe* changed to *Mione* (25/11/1897) to *Iolanthe* (6.10.1899) to *Amalthaea* (22/10/1906)

Steam Screw

Port, Port No. and Year: London, 72/1890 27/3/1890
105/1881 3/8/1881

Built At: Leith 1881
Ramage & Ferguson, Leith

Number of Decks: One
Number of Masts: Three
Rigged: Schooner
Stern: Square
Build: Carvel and Clencher
Galleries: None
Head: Female Bust
Framework and Plating: Iron

Length (ft): 189 3/10
Breadth (ft): 27 1/10
Depth (ft): 15

Register Tonnage: 260 90
100 altered *to* 204.08 on 13/3/1891

Engine Room Tonnage: 150.97
Length of Engine Room (ft): 35 1/10
No. of Engines: Two
Combined Power: (estimated HP) 110

Masters (where recorded): Thomas Williams – 3/8/1881
Francis R Harris – 6/6/1882
John William Creaghe – 27/3/1890

Owner: (i) Thomas James Waller of 60 Holland Park, Middlesex, Esq., address altered to South Villa, Regents Park, London; (ii) Sir Donald Currie KCMG, MP, Hyde Park Place, London – 64 shares

Closing Entry (i): Registry closed 24.7/1888 – vessel sold to a citizen of the United States of America

Re-purchased and registered in London No. 72/1890;

Closing Entry (ii): Registry closed 24/8/1920 – vessel totally wrecked at Stornoway on the 1st January 1919 in name of *Iolaire* while in Admiralty service.

Appendix III
THE LOST AND THE SAVED
1 JANUARY 1919

HER OFFICERS AND CREW

Commander Richard Gordon William Mason
75 Main Road, Handsworth, near Sheffield
Son of Robert and Julia Mason;
husband to Lucy Lavinia
Service: Royal Naval Reserve
Age: 44
Body never recovered.

Lieutenant Leonard Edmund Cotter
Cranmore House, Park Road, Cowes, Isle of Wight
Husband to Margaret Eleanor
Service: Royal Naval Reserve
Age: 49
Body never recovered.

Sub-Lieutenant (E) Charles Ritchie Rankin
Penzance
Son of John and Anne Rankin
Service: Royal Naval Reserve
Age: 30
Buried at Penzance.

2nd Engineer John Hern
13 Roxburgh Street, Sunderland
Son of John and Ann Isa Hern
Service: Mercantile Marine
Age: 26
Buried at Sunderland.

Deckhand Ernest Leggett
43 South Street, Emsworth
Service: Mercantile Marine Reserves
Age: 41
Buried at Havant.

Leading Deckhand Charles M. Dewsbury
1 Rodney Road, Great Yarmouth
Service: Mercantile Marine
Age: 33
Buried at Sandwick.

Signal Boy David MacDonald
53 Virginia Street, Aberdeen
Son of John and Mary MacDonald
Service: Royal Naval Reserve
Service number: 1265/SB
Age: 17 (youngest casualty)
Buried at Sandwick.

Assistant Steward Alfred S. Taylor
205 Gloucester Terrace, London W2
Service: Mercantile Marine
Age: 52

Buried at Sandwick.

Chief Petty Officer / Cook Alfred William Henley
Isle of Wight
Son of Richard and Urina Henley;
husband to Elizabeth
Service: RN
Service number: MFA/2932
Age: 45

Buried at St Helens.

Greaser Thomas Edward H. Harris
Portsmouth
Son of Sam and Emma Harris
Service: Mercantile Marine Reserves
Age: 27

Buried at Portsmouth.

Fireman Joseph W. George
Newcastle-upon-Tyne
Son of Sam and Margaret George;
husband to Jane
Service: Mercantile Marine
Age: 43

Buried at Newcastle.

Deckhand Harold Moore
8 Myrtle Road, Southend-on-Sea
Service: Mercantile Marine Reserves
Age: 30

Buried at Southend.

Seaman David Ramsay
Auchterarder
Husband to Martha
Service: Royal Naval Reserve
Service number: 1350/D
Age: 50

Buried at Auchterarder.

Deckhand William Joseph John Stanley
75 Calvert Road, East Greenwich
Son of William and Lucy Stanley
Service: Mercantile Marine Reserves
Age: 19

Buried at Greenwich.

Carpenter Frederick Charles McCarthy
14 Hope Street, West Hartlepool
Son of John and Jane McCarthy
Service: Mercantile Marine Reserves
Age: 34

Body never recovered.

Trimmer / Cook Ernest Ainsworth Brown
193 Welholme Road, Grimsby
Son of Stephen and Martha Brown
Service: Royal Naval Reserve
Service number: 902/TC
Age: 22

Body never recovered.

L/Vic Asst Albert Richard Matthews
27 Arthur Road, Holloway, London, N7
Son of William and Annie Matthews
Service: RN
Age: 22

Buried at Islington.

Steward Frank Humphrey
Southampton
Son of Adolphus and Eliza Humphrey
Service: Mercantile Marine
Ager: 41

Buried at Southampton.

Deckhand Henry Orley Mariner
217 Queen's Road, Buckland, Portsmouth
Son of Charles and Minnie Mariner
Service: Mercantile Marine Reserves
Service number: 821982
Age: 25

Body never recovered.

Private Herbert William Head
Church Hill, Monks Eleigh, Suffolk
Son of Catherine Head, engaged to be
married
Service: RMLI
Service number: PO/11997
Age: 35
Body never recovered.

Fireman Ernest Reginald Adams
Cardiff
Service: Royal Naval Reserve
Survived.

Fireman Griffith Ramsay
Cardiff
Service: Royal Naval Reserve
Survived.

Deckhand James MacLean
Campbeltown
Service: Royal Naval Reserve
Age: 39
Survived.

Trimmer Arthur Topham
Hull
Service: Royal Naval Reserve
Service number: 4165
Survived.

Deckhand J.F. Wilder
Residence unknown
Service: Royal Naval Reserve
Survived.

Able Seaman John MacLellan
Residence unknown
Service: Mercantile Marine Reserve
Survived.

Telegrapher Leonard Welch
Malvern
Service: RN
Age: 23
Survived.

ISLE OF LEWIS, BY CIVIL REGISTRATION DISTRICT
PARISH OF BARVAS
NESS

CROSS

Seaman Alexander Morrison
4 Cross
Son of John and Effie Morrison, *née*
MacLennan
Service: Royal Naval Reserve,
HMS *Pembroke*
Service number: 3207/C
Age: 44
Survived. Died May 1962.

Engine Room Artificer Murdo MacFarlane
24 Cross, Post Office
Son of Murdo and Elizabeth MacFarlane,
née MacPherson
Service: RN, HMS *Royal Sovereign*
Survived. Later emigrated to America.

Deckhand Norman MacKenzie
36 Cross
Son of Donald and Margaret MacKenzie,
née MacFarlane
Service: Royal Naval Reserve,
HMT *Conway*
Service number: 20010/DA
Age: 18
Survived – revived in temporary mortuary.
Died 14 July 1954.

DELL

Seaman Angus Gillies
35 South Dell
Son of Alexander and Isabella Gillies, *née*
MacKenzie
Service: Royal Naval Reserve,
HMY *Seahorse*
Service number: 4502/A
Age: 30
Buried at Swainbost.

John Murray
6 South Dell
Husband to Catherine Stewart
Service: Royal Naval Reserve
Age: 31
Survived. Had moved to Back on his marriage. Died 1966.

EORODALE

Deckhand Murdo Campbell
4 Eorodale
Son of Donald and Catherine Campbell, *née* MacLean
Service: Royal Naval Reserve,
HMT *Matthew Flynn*
Service number: 20536/DA
Age: 19
Buried at Swainbost.

Deckhand John MacLeod
13 Eorodale
Son of Donald and Mary MacLeod, *née* MacDonald
Service: Royal Naval Reserve,
HMS *Attentive III*
Service number: 15739/DA
Age: 20
Buried at Swainbost.

Engine Room Artificer Malcolm MacDonald
1 Eorodale
Service: Royal Naval Reserve, HMS *Vivid*
Service number: SB/1542
Survived. Later emigrated overseas.

EOROPIE

Deckhand Angus Morrison
10 Eoropie
Son of John and Catherine Morrison, *née* MacLeod
Service: Royal Naval Reserve,
HMS *Implacable*
Service number: 14310/DA
Age: 20
Body never recovered.

FIVEPENNY

Deckhand Donald MacLeod
5 Fivepenny
Son of Angus and Margaret MacLeod
Service: Royal Naval Reserve
Age: 28
Body never recovered.

Signaller William MacKay
7 Fivepenny
Son of William and Mary MacKay
Service: RNVR, HMS *Vivid*
Service number: Z/8218
Age: 26
Buried at Swainbost.

Deckhand Donald Morrison
11 Fivepenny
Son of Donald and Gormelia Morrison, *née* Campbell
Service: Royal Naval Reserve,
HMT *Sir Mark Sykes*
Service number: 11859/DA
Age: 27
Buried at Swainbost.

HABOST

Seaman Roderick Morrison
2 Back Street, Habost
Son of Malcolm and Margaret Morrison;
husband to Gormilia
Service: Royal Naval Reserve, HMS *Ganges*
Service number: 1750/CH
Age: 44
Buried at Swainbost.

Deckhand Donald Murray
11 Habost
Son of John and Margaret Murray, *née*
MacKenzie
Service: Royal Naval Reserve,
HMT *Joseph Burgin*
Service number: 19804/DA
Age: 23
Buried at Swainbost.

Deckhand Donald MacRitchie
34 Habost
Son of Finlay and Catherine MacRitchie,
née Morrison
Service: Royal Naval Reserve,
HMT *Scarboro*
Service number: 13258/DA
Age: 21
Buried at Swainbost.

Deckhand Alexander John Campbell
41 Habost
Son of John and Isabella Campbell, *née*
MacKay (North Tolsta)
Service: Royal Naval Reserve,
HMS *Venerable*
Service number: 11999/DA
Age: 22
Body never recovered.

KNOCKAIRD

Petty Officer (1st Class) Angus Morrison
7 Knockaird
Son of Donald and Jessie Morrison, *née*
Gunn (1861–1960; last *Iolaire* mother)
Service: Royal Naval Reserve, HMS *Vivid*
Service number: 5306/A
Age: 32
Body never recovered.

Deckhand John Morrison
12 Knockaird
Son of Norman and Margaret Morrison,
née Smith
Service: Royal Naval Reserve
Service number: 21746/DA
Age: 18
Buried at Swainbost.

Deckhand Donald Morrison
7 Knockaird
Service: Royal Naval Reserve, Trawler
Division, Portland
Service number: 17296/DA
Age: 20
Survived after clinging all night to mainmast.
Died 16 July 1990, aged 92.

LIONEL

Seaman Norman Morrison
17 Lionel
Son of Donald and Jessie Morrison, *née*
Campbell
Service: Royal Naval Reserve, HMT *Urka*
Service number: 12088/DA
Age: 21
Buried at Swainbost.

Seaman Angus Campbell
31 Lionel
Son of Donald and Catherine Campbell;
husband to Susy Ann Morrison
(Lochmaddy)
Service: Royal Naval Reserve,
HMS *Excellent*
Service number: 3590/C
Age: 44
Body never recovered.

Seaman John Murray
36 Lionel
Son of Norman and Mary Murray, *née*
Gunn; husband to Dolina Morrison (died
1967)
Service: Royal Naval Reserve,
HMS *Pembroke*
Service number: 2061/C
Age: 45
Buried at Swainbost.

PORT OF NESS

Deckhand Angus MacDonald
3 Port of Ness
Son of Angus and Margaret MacDonald,
née Campbell
Service: Royal Naval Reserve,
HMD *Primrose*
Service number: 2597/SD
Age: 24
*Buried at Crossbost. Last body to be recovered,
summer 1919.*

Seaman John Finlay MacLeod
4 Port of Ness
Son of Murdo and Catherine MacLeod, *née*
MacKay (Achmore)
Service: Royal Naval Reserve,
HMS *Ganges II*
Service number: 16774/DA
Age: 30
Survived. The hero of the Iolaire – *died 21
December 1978.*

SKIGERSTA

Seaman John MacDonald
10 Skigersta
Son of John and Christina MacDonald, *née*
MacLean; husband to Jessie Finlayson (m.
18/11/1918)
Service: Royal Naval Reserve, HMY *Seahorse*
Service number: 4490/A
Age: 32
Buried at Swainbost.

Deckhand Murdo Morrison
8 Skigersta
Service: Royal Naval Reserve, HMT *Horatio*
Service number: 16322/DA
Survived.

SWAINBOST

Seaman Donald MacDonald
13 Swainbost
Son of John and Mary MacDonald, *née*
Murray; brother of Murdo (below)
Service: Royal Naval Reserve, SS *Mandala*
Service number: 5351/A
Age: 27
Body never recovered.

Deckhand Murdo MacDonald
13 Swainbost
Son of John and Mary MacDonald, *née*
Murray; brother of Donald (above)
Service: Royal Naval Reserve, HMD *Aspire*
Service number: 11997/DA
Age: 21
Buried at Swainbost.

Seaman Malcolm Thomson
14 Swainbost
Son of John and Christina Thomson, *née*
MacLean
Service: Royal Naval Reserve,
HMS *Redoubtable*
Service number: 4557/A
Age: 27
Buried at Swainbost.

Ordinary Seaman Malcolm MacLeod
28 Swainbost
Son of Murdo and Margaret MacLeod
Service: RN, HMS *Maidstone*
Service number: J/65506
Age: 21
Body never recovered.

Deckhand Angus MacRitchie
38 Swainbost
Son of Donald and Effie (Henrietta)
MacRitchie, *née* MacKenzie
Service: Royal Naval Reserve, HMT *Hero*
Service number: 16522/DA
Age: 20
Buried at Swainbost.

WEST SIDE

ARNOL

Seaman Norman MacLeod
13 Arnol
Son of Norman MacLeod; husband to
Christina MacLeod (died 8 September 1933,
aged 49)
Service: Royal Naval Reserve
Service number: 3343/C
Age: 36
*Buried at Bragar. His younger son, Rev. John
MacLeod (1918–95), was a noted bard.*

Seaman Kenneth MacPhail
24 Arnol
Son of Malcolm and Catherine MacPhail
Service: Royal Naval Reserve,
HMS *Pembroke*
Service number: 3320/A
Age: 28
Buried at Bragar.

Seaman Donald MacDonald
35 Arnol
Son of Donald and Henrietta MacDonald
Service: Royal Naval Reserve, HMS *Vernon*
Service number: 5296/A
Age: 23
Body never recovered.

Seaman Angus MacIver
45 Arnol
Son of Norman and Catherine MacIver;
husband to Catherine
Service: Royal Naval Reserve
Service number: 1855/C
Age: 36
Buried at Bragar.

Leading Seaman Norman MacIver
21 Arnol
Service: Royal Naval Reserve,
HMT *Magnet III*
Service number: 2621/SD
Age: 24
Survived.

BARVAS

Seaman Donald MacLeod
20 Lower Barvas
Son of Murdo and Catherine MacLeod
Service: Royal Naval Reserve, SS *Saxonia*
Service number: 3553/B
Age: 33
Buried at Barvas.

BORVE

Seaman Murdo MacDonald
15 Borve
Son of Murdo and Mary Ann MacDonald,
née Graham
Service: Royal Naval Reserve,
HMS *Pembroke*
Service number: 9534/A
Age: 18
Body never recovered.

Deckhand Murdo Graham
8 Borve,
Service: Royal Naval Reserve,
HMS *Excellent*
Service number: 13729/DA
Survived. Later emigrated to New Zealand.

Deckhand Roderick Graham
29 Borve
Son of Roderick and Gormelia Graham,
née Morrison
Service: Royal Naval Reserve,
HMS *Attentive III*
Survived. Died in the 1920s.

Seaman Angus Morrison
41 Borve,
Service: Royal Naval Reserve, HMS *Canopus*
Service number: 3363/A
Survived.

SHADER

Seaman Norman Martin
8 Lower Shader
Son of John and Margaret Martin;
husband to Annie
Service: Royal Naval Reserve, HMS *Victory*
Service number: 3397/C
Age: 42
Buried at Barvas.

Deckhand John MacDonald
25 Lower Shader
Son of John and Mary MacDonald, *née*
MacLeod
Service: Royal Naval Reserve,
HMD *Boy George III*
Service number: 19654/DA
Age: 32
Buried at Barvas.

Seaman Angus MacLeay
34 Lower Shader
Son of John and Annie MacLeay;
husband to Kate Ann Martin
Service: Royal Naval Reserve,
HMS *Emperor of India*
Service number: 3689/B
Age: 38
Buried at Barvas.

Leading Deckhand Malcolm Matheson
10 Upper Shader
Son of Malcolm and Catherine Matheson
Service: Royal Naval Reserve, HMT *Ireland*
Service number: 11907/DA
Age: 27
Buried at Barvas

Deckhand Angus Morrison
31 Upper Shader
Son of Malcolm and Chirsty Morrison;
husband to Katy Ann
Service: Royal Naval Reserve, HMT *St Ayles*
Service number: 12126/DA
Age: 20
Buried at Barvas.

Deckhand Donald Martin
33 Lower Shader
Service: Royal Naval Reserve
Survived. Later emigrated to Canada with his brother.

CIVIL PARISH OF CARLOWAY
CARLOWAY DISTRICT
BORROWSTON
Seaman Donald MacPhail
11 Borrowston
Son of Donald and Marion MacPhail;
husband to Marion
Service: Royal Naval Reserve,
HMS *Pembroke*
Service number: 2222/D
Age: 44
Buried at Dalmore.

Seaman John MacLean
12 Borrowston
Service: Royal Naval Reserve, HMML 511
Service number: 4139/S
Age: 22
Survived. Died 1983, aged 87.

Seaman Malcolm MacLeod
16 Borrowston
Service: Royal Naval Reserve,
HMS *Tedworth*
Service number: 2410/C
Survived.

BREASCLETE
Seaman Malcolm MacIver
40 Breasclete
Son of Neil and Christy MacIver
Service: Royal Naval Reserve, HMS *Victory*
Service number: 2778/A
Age: 35
Buried at Dalmore.

Leading Deckhand Donald MacDonald
44 Breasclete
Son of Alexander and Effie Macdonald
Service: Royal Naval Reserve,
HMS *Pembroke*
Service number: 12453/DA
Age: 20
Buried at Dalmore.

DOUNE
Seaman John MacPhail
13 Doune, Carloway
Service: Royal Naval Reserve, HMS *Arab*
Service number: 3428/B
Age: 41
Survived.

Deckhand Angus Macphail
13 Doune, Carloway
Service: Royal Naval Reserve,
HMS *Dreel Castle*
Service number: 3167/SD
Survived.

GARENIN
Seaman Murdo MacKenzie
15 Garenin
Son of John and Christy MacKenzie;
husband to Christina
Service: Royal Naval Reserve,
HMS *President*
Service number: 2793/C
Age: 36
Buried at Dalmore.

KIRIVICK
Deckhand John MacAskill
3 Kirivick
Son of Donald and Maggie MacAskill,
Service: Royal Naval Reserve,
HMT *Sir John Fitzgerald*
Service number: 14567/DA
Age: 20
Buried at Dalmore.

Deckhand John MacLeod
6 Kirivick
Son of Malcolm and Mary MacLeod
Service: Royal Naval Reserve,
HMS *Implacable*
Service number: 19980/DA
Age: 19
Buried at Dalmore.

Deckhand Donald MacArthur
12 Kirivick
Son of Malcolm and Isabella MacArthur;
husband to Jane
Service: Royal Naval Reserve,
HMS *Implacable*
Service number: 4443/SD
Age: 35
Body never recovered.

KNOCK (Carloway)

Deckhand Norman MacPhail
Knock, Carloway
Son of Norman and Ann MacPhail;
husband to Maggie
Service: Royal Naval Reserve,
HMS *Venerable*
Service number: 14663/DA
Age: 37
Buried at Dalmore.

George Morrison
Knock, Carloway
Service: Royal Naval Reserve
Survived.

TOLSTA CHAOLAIS

Deckhand Alexander Angus MacLeod
28 Tolsta Chaolais
Son of Alexander and Annie MacLeod
Service: Royal Naval Reserve, HMS *Idaho*
Service number: 3455/SD
Age: 20
Body never recovered.

Seaman Alexander MacPhail
18 Tolsta Chaolais
Service: Royal Naval Reserve, HMS *Victory*
Service number: 3364/A
Survived.

WEST SIDE

BRAGAR

Seaman Murdo MacLean
6 South Bragar
Son of Angus and Annie MacLean;
husband to Effie Campbell; brother of John
(17 South Bragar)
Service: Royal Naval Reserve, HMS *Victory*
Service number: 1903/D
Age: 42
Body never recovered.

Seaman Malcolm MacLean
10 South Bragar
Son of Murdo and Catherine MacLean
Service: Royal Naval Reserve, HMS *Vernon*
Service number: 2679/C
Age: 48
Body never recovered.

Seaman John MacLean
17 South Bragar
Son of Angus and Annie MacLean;
brother of Murdo (6 South Bragar)
Service: Royal Naval Reserve
Service number: 4280/B
Age: 37
Buried at Bragar.

Deckhand John Murray
30 South Bragar
Service: Royal Naval Reserve,
Mercantile Marine
Service number: 4689/SD
Age: 25
Body never recovered.

Seaman Malcolm MacKay
36 South Bragar
Husband to Kate MacLeod (Brue)
Service: Royal Naval Reserve,
HMS *Imperieuse*
Service number: 2613/C
Age: 38
Buried at Bragar.

Seaman Malcolm MacDonald
57 South Bragar
Son of Angus and Annie MacDonald;
husband to (i) Mary (ii) Isabella
Service: Royal Naval Reserve, HMS *Victory*
Service number: 1874/C
Age: 44
Body never recovered.

Deckhand Murdo MacKay
7 North Bragar
Son of Norman and Peggy MacKay
Service: Royal Naval Reserve,
HMS *Shikari II*
Service number: 16652/DA
Age: 21
Buried at Bragar.

Deckhand Murdo MacDonald
3 Fevig, South Bragar
Son of Murdo MacDonald
Service: Royal Naval Reserve,
HMS *Venerable*
Service number: 20516/DA
Age: 18
Buried at Bragar.

Deckhand Donald Murray
46 South Bragar
Service: Royal Naval Reserve, HMS *Olympia*
Service number: 17442/DA
Age: 19
Survived. Died 26 October 1981, aged 82.

SHAWBOST

Deckhand Donald MacLeod
5 South Shawbost
Son of Norman and Effie MacLeod
Service: Royal Naval Reserve
Service number: 4944/SD
Age: 20
Buried at Bragar.

Seaman John Smith
11 South Shawbost
Son of John and Mary Smith;
husband to Kate
Service: Royal Naval Reserve,
HMS *Duchess of Devonshire*
Service number: 3516/B
Age: 35
Buried at Bragar.

Seaman Angus MacLeod
11 South Shawbost
Son of Peter and Catherine MacLeod
Service: Royal Naval Reserve, HMS *Vernon*
Service number: 2020/D
Age: 51 (oldest casualty)
Body never recovered.

Ordinary Seaman Roderick Murray
25 South Shawbost
Son of Donald and Mary Murray
Service: RN, HMS *Roxburgh*
Service number: J/86329
Age: 19
Buried at Bragar

Seaman Donald William Gillies
30 South Shawbost
Son of Angus and Mary Gillies
Service: Royal Naval Reserve
Service number: 3309/A
Age: 28
Buried at Bragar.

Petty Officer Donald Murray
43 South Shawbost
Son of John and Mary Murray;
husband to Murdina
Service: Royal Naval Reserve,
HMS *Imperieuse*
Service number: 2811/B
Age: 40
Buried at Bragar.

Deckhand Donald Nicolson
10 North Shawbost
Son of Kenneth and Peggy Nicolson;
husband to Catherine
Service: Royal Naval Reserve, HMT *Calera*
Service number: 8955/DA
Age: 50
Buried at Bragar.

Deckhand Malcolm MacLeod
32 North Shawbost
Son of Murdo and Catherine MacLeod
Service: Royal Naval Reserve, HMT *Sabreur*
Service number: 20774/DA
Age: 18
Buried at Bragar.

Deckhand Donald MacLeod
38 North Shawbost
Son of Peter Angus and Christina
MacLeod
Service: Royal Naval Reserve, HMD *Beatrice*
Service number: 4125/SD
Age: 20
Buried at Bragar.

Seaman Murdo Morrison
31 North Shawbost
Service: Royal Naval Reserve,
HMS *Pembroke*
Service number: 9523/A
Survived.

Seaman Angus Peter Morrison
46 North Shawbost
Service: Royal Naval Reserve, Mine School,
Portsmouth
Service number: 4586/A
Survived.

PARISH OF LOCHS
KINLOCH DISTRICT
BALALLAN

Deckhand Malcolm MacLeod
18 Balallan
Son of Malcolm and Johanna MacLeod
Service: Royal Naval Reserve, HMS *Idaho*
Service number: 14384/DA
Age: 20
Body never recovered.

Deckhand Malcolm Martin
21 Balallan
Son of Donald and Christina Martin
Service: Royal Naval Reserve,
HMS *Pembroke*
Service number: 12067/DA
Age: 38
Buried at Laxay.

LAXAY

Deckhand John MacLeod
25 Laxay
Son of Colin and Catherine MacLeod
Service: Royal Naval Reserve,
Mercantile Marine 411
Service number: 3607/SD
Age: 25
Buried at Laxay.

Leading Seaman Malcolm MacLeod
Laxay
Service: Royal Naval Reserve
Age: 48
Survived. Died 27 October 1953, aged 83.

North Lochs District

Achmore

Leading Seaman Donald Smith
5 Achmore
Son of John and Christy Smith
Service: Royal Naval Reserve
Service number: 4173/A
Age: 26.
Body never recovered.

Deckhand Allan MacKay
1 Achmore
Service: Royal Naval Reserve
Survived.

Crossbost

Deckhand Donald MacLeod
2 Crossbost
Son of Malcolm MacLeod; uncle of
Malcolm, below
Service: Royal Naval Reserve
Service number: unknown
Age: 51
Body never recovered.

Deckhand Malcolm MacLeod
2 Crossbost (later resident at 42 Keith
Street, Stornoway)
Son of Malcolm MacLeod;
nephew of Donald, above
Service: Royal Naval Reserve,
HMS *Wallington*
Service number: 12081/DA
Age: 20
Buried at Sandwick.

Able Seaman John MacLeod
13 Crossbost
Son of Malcolm and Lillias MacLeod
Service: Mercantile Marine Reserves,
HMS *Rose*
Service number: 974908
Age: 20
Buried at Crossbost.

Murdo MacKenzie
Crossbost
Service: Royal Naval Reserve
Survived.

Grimshader

Able Seaman Malcolm MacLeod
3 Grimshader
Son of John and Mary MacLeod
Service: Mercantile Marine Reserves,
HMT *Agnes Nutten*
Service number: 973832
Age: 18
Buried at Crossbost.

Seaman John MacAulay
11 Grimshader
Son of Angus and Annabella MacAulay;
husband to Alexina
Service: Royal Naval Reserve,
HMS *Imperieuse*
Service number: 1567/C
Age: 46
Buried at Crossbost.

Leurbost

Seaman Alexander Angus MacKenzie
11 Leurbost
Son of Alexander and Isabella MacKenzie
Service: Royal Naval Reserve,
HMT *Roman Empire*
Service number: 1663/C
Age: 40
Buried at Crossbost.

Deckhand Alexander MacKenzie
16 Leurbost
Son of Donald and Mary MacKenzie;
brother of John
Service: Royal Naval Reserve, HMD *Mary*
Service number: 12080/DA
Age: 20
Buried at Crossbost.

Seaman John MacKenzie
16 Leurbost
Son of Donald and Mary MacKenzie,
brother of Alexander
Service: Royal Naval Reserve,
HMS *Emperor of India*
Service number: 32741/A
Age: 30
Buried at Crossbost.

Seaman Kenneth Smith
28B Leurbost
Son of Murdo and Mary Smith;
husband to Annie
Service: Royal Naval Reserve,
HMS *Pembroke*
Service number: 1956/D
Age: 41
Body never recovered.

Seaman Donald Smith
34 Leurbost
Son of Roderick Smith
Service: Royal Naval Reserve,
HMS *Emperor of India*
Service number: 2972/A
Age: 27
Buried at Crossbost.

Seaman Donald MacLean
35 Leurbost
Son of George and Isabella MacLean;
husband to Annie
Service: Royal Naval Reserve, HMS *Colleen*
Service number: 2820/B
Age: 50
Buried at Crossbost.

Seaman Roderick John MacDonald
36 Leurbost
Son of Murdo and Maggie MacDonald
Service: Royal Naval Reserve,
HMS *Wallington*
Service number: 2966/A
Age: 27
Buried at Crossbost.

Deckhand Murdo MacLean
39 Leurbost
Son of Alexander and Marion MacLean
Service: Mercantile Marine, HMS *Snipe*
Service number: 974252
Age: 31
Buried at Crossbost.

Seaman Angus MacDonald
42 Leurbost
Son of Allan and Mary MacDonald;
husband to Mary
Service: Royal Naval Reserve,
HMS *Imperieuse*
Service number: 1830/D
Age: 45
Buried at Crossbost.

Deckhand Angus MacLeod
46 Leurbost
Son of Angus and Marion MacLeod
Service: Mercantile Marine, HMS *Snipe*
Service number: 973997
Age: 27
Buried at Crossbost.

John MacDonald
3 Leurbost
Service: Royal Naval Reserve
Survived.

Deckhand Alexander MacKenzie
10 Leurbost
Service: Royal Naval Reserve, HMS *Valorous*
Survived.

Neil MacKenzie
14 Leurbost
Service: Royal Naval Reserve
Survived.

Seaman Archibald Ross
29 Leurbost
Service: Royal Naval Reserve, HMS *Vernon*
Age: 50
Survived.

Leading Seaman Alexander Smith
53 Leurbost
Service: Royal Naval Reserve,
Naval Base, Granton
Service number: 3269/A
Age: 26
Survived.

RANISH

Seaman Malcolm Nicolson
20 Ranish
Son of John and Marion Nicolson
Service: Royal Naval Reserve, HMS *Magpie*
Service number: 5396/A
Age: 23
Body never recovered.

Seaman Alexander Angus MacLeod
21 Ranish
Son of Alexander and Isabella MacLeod;
husband to Marion
Service: Royal Naval Reserve,
HMS *Imperieuse*
Service number: 2745/A
Age: 44
Body never recovered.

Leading Deckhand Donald MacDonald
23B Ranish
Son of John and Margaret MacDonald
Service: Royal Naval Reserve,
HMS *Dreel Castle*
Service number: 3516/SD
Age: 27
Buried at Crossbost.

Mate John MacLeod
31 Ranish
Son of William and Catherine MacLeod;
husband to Maggie
Service: Mercantile Marine, HMD *Cornrig*
Service number: unknown
Age: 39
Buried at Crossbost.

Deckhand Donald MacAulay
41 Ranish
Son of Malcolm and Margaret MacAulay
Service: Royal Naval Reserve, HMS *Gunner*
Service number: 2576/A
Age: 32
Body never recovered.

Seaman Allan MacLeod
43 Ranish
Son of Leod and Joan MacLeod
Service: Royal Naval Reserve,
HMS *President III*
Service number: 4661/A
Age: 25
Body never recovered.

Deckhand Colin MacDonald
3 Ranish
Service: Royal Naval Reserve, HMS *Etrarian*
Service number: 12785/DA
Survived.

John MacKinnon
9 Ranish
Service: Royal Naval Reserve
Survived.

Leading Seaman Donald MacRae
35 Ranish
Service: Royal Naval Reserve, HMS *Victory*
Service number: 4919/A
Age: 27
Survived.

Seaman Roderick Nicolson
36 Ranish
Service: Royal Naval Reserve, SS *Zaria*
Service number: 3160/A
Survived.

Seaman John Montgomery
Ranish
Service: Royal Naval Reserve, MLX 74
Service number: 3708/B
Age: 33
Survived.

SOUTH LOCHS DISTRICT

CAVERSTA

Deckhand Angus MacKinnon
4 Caversta
Son of Donald and Margaret MacKinnon
Service: Royal Naval Reserve,
HMS *Attentive III*
Service number: 2615/SD
Age: 22
Body never recovered.

CROMORE

Seaman George MacArthur
10 Cromore
Service: Royal Naval Reserve, HMS *Victory*
Survived.

Seaman Murdo Alasdair MacArthur
10 Cromore
Service: Royal Naval Reserve, HMS
Pembroke
Survived.

Deckhand Donald MacDonald
23 Cromore
Service: Royal Naval Reserve, SS *Zaria*
Service number: 16362/DA
Survived.

GARYVARD

Deckhand Alexander MacLeod
1 Garyvard
Son of John and Annabella MacLeod
Service: Royal Naval Reserve, HMS *Vivid*
Service number: 17041/DA
Age: 26
Body never recovered.

Deckhand Angus Montgomery
2 Garyvard
Son of Kenneth and Peggy Montgomery;
husband to Mary Ann
Service: Royal Naval Reserve, HMT *Unity*
Service number: 1445/SD
Age: 44
Buried at Laxay.

GRAVIR

Deckhand Donald MacAskill
9 Gravir
Son of Donald and Margaret MacAskill
Service: Royal Naval Reserve, HMS *Gunner*
Service number: 21143/DA
Age: 33
Body never recovered.

Deckhand Allan MacAskill
9 Gravir
Service: Royal Naval Reserve,
HMD *Boy Scout*
Survived.

LEMREWAY

Deckhand David MacInnes
2 Lemreway
Son of Peter and Annie MacInnes
Service: Royal Naval Reserve,
HMT *Daniel Henry*
Service number: 19531/DA
Age: 19
Buried at Gravir.

Seaman Malcolm MacInnes
2 Lemreway
Son of Murdo and Mary MacInnes
Service: Royal Naval Reserve, HMS *Dublin*
Service number: 8896/A
Age: 26
Body never recovered.

Deckhand Murdo Ferguson
3 Lemreway
Son of Murdo and Christina Ferguson;
husband to Betsy
Service: Royal Naval Reserve,
HMT *Morgan Jones*
Service number: 258/SD
Age: 48
Buried at Gravir.

Able Seaman Neil Nicolson
20 Lemreway
Service: Royal Naval Reserve, HMD
Returdo; survived sinkings of HMS *Avenger*
and HMS *Campaign*
Service number: 88951
Age: 23
Served throughout Second World War; died
27 June 1992, aged ninety-six – the last
survivor of the Iolaire *disaster.*

MARVIG

Seaman John MacKenzie
13 Marvig
Son of Alexander and Maggie MacKenzie
Service: Royal Naval Reserve,
HMS *Albemarle*
Service number: 5397/A
Age: 27
Body never recovered.

Able Seaman Roderick Finlayson
8 Marvig
Service: Mercantile Marine Reserves,
HMS *Eaglet*
Survived.

PARISH OF STORNOWAY

BACK DISTRICT

BACK

Deckhand Donald MacDonald
11 Back
Son of Murdo and Catherine MacDonald
Service: Royal Naval Reserve, HMT *Santora*
Service number: 17809/DA
Age: 19
Buried at Gress.

Deckhand William John 'Robert' Murray
Well Cottage, Lighthill (Back)
Son of Angus and Agnes Murray
Service: Royal Naval Reserve, HMS *Pactolus*
Service number: 11935/DA
Age: 21
Buried at Gress.

COLL

Deckhand William MacLeod
8 Coll
Son of Angus and Margaret MacLeod
Service: Royal Naval Reserve,
HMS *Dreel Castle*
Service number: 14603/DA
Age: 20
Body never recovered.

Leading Seaman John Morrison
10 Coll
Son of John Morrison;
husband to Catherine
Service: Royal Naval Reserve, SS *Norwood*
Service number: 30260(CH)
Age: 44
Buried at Gress.

Able Seaman Murdo MacLeod
30 Coll
Son of Murdo and Catherine MacLeod;
husband to Ann MacLeod
Service: RN, HMS *Revenge*
Service number: J/76479
Age: 27
Body never recovered.

Able Seaman Alexander Beaton
40 Coll
Son of James and Isabella Beaton
Service: Mercantile Marine Reserves,
HMS *Rose III*
Service number: 968068
Age: 28
Buried at Gress.

Deckhand Alexander John MacLeod
63 Coll
Service: Royal Naval Reserve,
HMD *Scotsman*
Service number: 11991/D
Age: 20
Survived. Died 8 January 1973, aged 74.

VATISKER

Leading Seaman Donald Campbell
3 Vatisker
Son of Norman and Margaret Campbell;
husband to Catherine; brother of
Alexander, below
Service: Royal Naval Reserve,
HMS *Pembroke*
Service number: 2058/C
Age: 50
Buried at Gress.

Leading Seaman Alexander Campbell
8 Vatisker
Son of Norman Campbell; husband to
Jessie; brother of Donald, above
Service: Royal Naval Reserve, HMS *Victory*
Service number: 2999/C
Age: 42
Body never recovered.

Seaman John MacAskill
Lighthill (Vatisker)
Son of Murdo and Ann MacAskill
Service: Royal Naval Reserve,
HMS *Redoubtable*
Service number: 3397/A
Age: 27
Buried at Gress.

POINT DISTRICT

AIGNISH

Deckhand Malcolm MacLeod
5 Aignish
Son of Donald and Catherine MacLeod
Service: Royal Naval Reserve, HMML 485
Service number: 4793/SD
Age: 19
Body never recovered.

Deckhand Malcolm MacIver
28 Aignish
Son of Malcolm and Catherine MacIver
Service: Royal Naval Reserve
Service number: 10235/DA
Age: 37
Buried at Aignish.

AIRD

Seaman Alexander MacKenzie
1 Aird, Point
Son of William and Mary MacKenzie;
husband to Marion
Service: Royal Naval Reserve
Service number: 3360/C
Age: 42
Buried at Aignish.

Seaman Alexander MacKenzie
5 Aird, Point
Son of Donald and Catherine MacKenzie;
husband to Margaret
Service: Royal Naval Reserve
Service number: 3892/B
Age: 41
Buried at Aignish.

Seaman Murdo MacLeod
10 Aird, Point
Son of Murdo and Margaret MacLeod;
husband to Anne
Service: Royal Naval Reserve
Service number: 4219/B
Age: 45
Buried at Aignish

BAYBLE

Seaman John Smith
17 Upper Bayble
Son of Angus and Isabella Smith;
husband to Annie
Service: Royal Naval Reserve,
HMS *Pembroke*
Service number: 8055A
Age: 46
Body never recovered.

Deckhand Donald MacKenzie
22 Upper Bayble
Son of Roderick and Catherine MacKenzie;
husband to Peggy
Service: Mercantile Marine Reserves
Service number: 967820
Age: 50
Body never recovered.

Seaman John MacLeod
43 Upper Bayble
Son of Murdo and Hannah MacLeod;
husband to Catherine
Service: Royal Naval Reserve,
HMS *Excellent*
Service number: 27361/B
Age: 42
Buried at Sandwick.

Seaman Malcolm MacMillan
Last address in Lewis: 51 Upper Bayble
Son of Malcolm and Christy MacMillan;
husband to Catherine
Service: Royal Naval Reserve, HMY *Iolaire*
Service number: 1848/D
Age: 46
Buried at Aignish.

Deckhand William MacDonald
44 New Park Bayble
Son of Donald and Catherine MacDonald
Service: Royal Naval Reserve, HMS *Surf*
Service number: 14360/DA
Age: 20
Buried at Aignish.

Mate John MacIver
19 Lower Bayble
Son of John and Margaret MacIver;
husband to Margaret
Service: Mercantile Marine Reserves
Age: 46
Body never recovered.

Leading Seaman Alexander MacDonald
28 Lower Bayble
Son of John and Bella MacDonald;
husband to Margaret
Service: Royal Naval Reserve, HMS *Nairn*
Service number: 2046/C
Age: 43
Body never recovered.

Deckhand Murdo MacIver
36 Lower Bayble
Son of Angus and Isabella MacIver;
husband to Annie
Service: Royal Naval Reserve,
HMT *Zena Dare*
Service number: 775/SD
Age: 49
Body never recovered.

John MacKenzie
51 Upper Bayble
Service: Royal Naval Reserve
Age: 35
Survived. Died 11 November 1955, aged 80.

Seaman Murdo Stewart
9 Lower Bayble
Service: Royal Naval Reserve
Survived.

Donald MacLeod
11 Lower Bayble
Service: Royal Naval Reserve
Survived.

Donald MacDonald
30 Geilar, Bayble
Service: Royal Naval Reserve – discharged
by 31 December 1918
Age: 38
Survived.

BROKER

Deckhand Alexander John MacLeod
3 Broker
Son of Murdo and Mary MacLeod
Service: Royal Naval Reserve
Service number: 20422/DA
Age: 18
Buried at Aignish.

GARRABOST

Deckhand Norman MacKenzie
1 Church St, Garrabost
Son of John and Mary MacKenzie
Service: Royal Naval Reserve, HMS *Victory*
Service number: 20072/DA
Age: 18
Buried at Ui – old Aignish churchyard.

Deckhand John MacLeod
30 Lower Garrabost
Son of Alexander and Catherine MacLeod;
husband to Maggie
Service: Mercantile Marine Reserves,
HMS *Victory*
Service number: 968097
Age: 37
Buried at Aignish.

KNOCK (Point)

Seaman Angus Crichton
12 Knock, Point
Son of Colin and Jessie Crichton;
husband to Mary
Service: Royal Naval Reserve, HMS *Ganges*
Service number: 2687/B
Age: 42
Body never recovered.

Seaman Donald Crichton
15 Knock, Point
Son of Alexander and Mary Crichton;
husband to Mary
Service: Royal Naval Reserve,
HMS *Pembroke*
Service number: 9066/A
Age: 23
Body never recovered.

Seaman Angus MacLeod
18 Knock, Point
Son of Torquil and Mary MacLeod
Service: Royal Naval Reserve
Service number: 4548/SD
Age: 23
Buried at Aignish.

PORTNAGURAN

Seaman Angus MacLeod
1 Portnaguran
Natural son of Margaret MacLeod,
4 Garrabost; husband to Margaret
Service: Royal Naval Reserve
Service number: 2808/C
Age: 42
Buried at Aignish.

Trimmer/Cook Norman MacLeod
10 Portnaguran
Son of Donald and Christy MacLeod
Service: Royal Naval Reserve,
HMS *Venerable*
Service number: 1186 TC
Age: 20
Buried at Aignish.

PORTVOLLER

Seaman John MacKenzie
5 Portvoller
Service: Royal Naval Reserve, HMS
Excellent; Whale Island Depot, Portsmouth
Service number: 4508/B
Age: 35
Survived. Witness at FAI.

Seaman Malcolm Martin
13 Portvoller
Service: Royal Naval Reserve
Service number: 2326C
Survived.

Seaman John MacKenzie
8 Portvoller
Service: Royal Naval Reserve,
HMS *Pembroke*
Service number: 3297/B
Survived.

SHESHADER

Seaman Donald MacDonald
5 Sheshader
Son of John and Christina MacDonald
Service: Royal Naval Reserve,
HMS *Muskerry*
Service number: 2373/C
Age: 41
Buried at Aignish.

Seaman Norman Montgomery
6 Sheshader
Son of Malcolm Montgomery;
husband to Isabella
Service: Royal Naval Reserve, HMT *Ariel II*
Service number: 3391/C
Age: 37
Body never recovered.

Deckhand Murdo MacAulay
7 Sheshader
Son of William and Kate MacAulay
Service: Royal Naval Reserve
Service number: 4004/SD
Age: 36
Buried at Aignish.

Seaman William Murray
11 Sheshader
Son of John and Matilda Murray
Service: Royal Naval Reserve,
HMS *Pembroke*
Service number: 6772/A
Age: 22
Body never recovered.

Deckhand Donald M. MacAulay
13 Sheshader
Son of John and Christina MacAulay
Service: Royal Naval Reserve
Service number: 4363/SD
Age: 19
Buried at Aignish.

Seaman Murdo MacKenzie
15 Sheshader
Son of Kenneth and Nancy MacKenzie;
husband to Dolina
Service: Royal Naval Reserve,
HMS *Pembroke*
Service number: 3122/B
Age: 47
Body never recovered.

Seaman John MacDonald major
20 Sheshader
Son of Kenneth and Catherine MacDonald
Service: Royal Naval Reserve,
HMS *Venerable*
Service number: 2558A
Age: 29
Buried at Aignish.

Seaman John MacDonald minor
20 Sheshader
Son of Kenneth and Catherine MacDonald;
brother of John, above
Service: Royal Naval Reserve,
HMS *Emperor of India*
Service number: 3074A
Age: 26
Buried at Aignish.

Deckhand Donald MacKay
22 Sheshader
Son of John and Matilda MacKay
Service: Royal Naval Reserve
Service number: 12090/DA
Age: 20
Buried at Aignish.

Leading Seaman Norman MacLeod
23 Sheshader
Son of Malcolm and Annie MacLeod
Service: Royal Naval Reserve
Service number: 4803/A
Age: 29
Buried at Aignish.

SHULISHADER

Seaman Donald MacAulay
1 Shulishader
Son of John and Annie MacAulay;
husband to Rachel
Service: Royal Naval Reserve,
HMS *Emperor of India*
Service number: 2065/C
Age: 39
Body never recovered.

Seaman Kenneth MacKenzie
4 Newlands, Shulishader
Son of Donald and Isabella MacKenzie
Service: Royal Naval Reserve
Service number: 3046/A
Age: 27
Buried at Aignish.

2nd Hand Angus MacKay
11 Shulishader
Son of Murdo and Catherine MacKay;
husband to Annie
Service: Royal Naval Reserve
Service number: 2258/D
Age: 41
Buried at Aignish.

Seaman Donald MacAskill
14 Shulishader
Son of Donald and Margaret MacAskill
Service: Royal Naval Reserve,
HMS *Sigismund*
Service number: 7041/A
Age: 21
Body never recovered.

Leading Seaman Alexander MacIver
19 Shulishader
Son of Angus and Mary MacIver;
husband to Mary
Service: Royal Naval Reserve
Service number: 1691/C
Age: 43

Buried at Aignish.

Leading Seaman John MacKay
7 Shulishader
Service: Royal Naval Reserve,
HMS *Emperor of India*
Service number: 1473C
Age: 48

*Survived. Witness at Court of Inquiry and
the FAI.*

Petty Officer Duncan MacAskill
14 Shulishader
Service: Royal Naval Volunteer Reserve
Service number: 2809

*Survived. Brother of Donald, above – had to
let go of him in the water to save his own life.
Later married and moved to Carloway; died
8 October 1935 aged forty-six, a year after the
loss of his 11-year-old son. Grandfather of Rev.
Duncan MacLeod, Knox Free Church, Perth.*

Seaman Angus MacAulay
Newlands, Shulishader
Service: Royal Naval Reserve, Longhope
Service number: 4060B

Survived.

Swordale

2nd Hand Roderick MacKenzie
5 Swordale
Son of Murdo and Ann MacKenzie;
husband to Ann MacDonald
Service: Royal Naval Reserve, HMT *Romilly*
Service number: 7246/DA
Age: 32

Body never recovered.

Seaman Murdo MacKay
16 Swordale
Son of Angus and Christy MacKay
Service: Royal Naval Reserve
Service number: 4511/B
Age: 31

Buried at Aignish.

Deckhand Alexander Campbell
26 Swordale
Son of James and Mary Campbell
Service: Royal Naval Reserve,
HMS *Venerable*
Service number: 13353/DA
Age:19

Body never recovered.

Seaman Kenneth MacLeod
28 Swordale
Service: Royal Naval Reserve, SS *Dalton*
Service number: 3138/B
Age: 40

*Survived. Witness at Court of Inquiry and
FAI.*

STORNOWAY DISTRICT

BRANAHUIE

Signalman Norman Buchanan
7 Branahuie
Service: Royal Naval Reserve,
HMD Sphinx II

Survived.

HOLM

Deckhand John MacDonald
10 Holm
Son of Alexander and Margaret
MacDonald
Service: Royal Naval Reserve, HMD *Genia*
Service number: 21078/DA
Age: 18

Buried at Sandwick.

NEWVALLEY

Deckhand Angus MacLeod
1 Newvalley
Son of Alexander MacLeod
Service: Royal Naval Reserve, HMT *Resmilo*
Service number: 19972/DA
Age: 18

Buried at Crossbost.

2nd Hand Alexander MacDonald
7 Newvalley
Son of Donald and Margaret MacDonald
Service: Royal Naval Reserve,
HMS *Pembroke*
Service number: 11708/DA
Age: 40

Body never recovered.

SANDWICK

Leading Deckhand John MacAskill
12 Lower Sandwick
Son of Kenneth and Mary MacAskill
Service: Royal Naval Reserve,
HMT *Thomas Booth*
Service number: 9635/DA
Age: 24

Buried at Sandwick.

STORNOWAY

Cooper 4th Class Donald MacRitchie
46 Keith Street, Stornoway
Natural son of Kirsty MacRitchie;
husband to (i) Jane MacIver (ii) Annie
Service: RN, HMS *Pembroke*
Service number: M/23885(CH)
Age: 29

Buried at Sandwick.

Signaller John Alex "Jack" MacAskill
75 Keith Street
Son of Hugh and Christina MacAskill
Service: Royal Naval Volunteer Reserve,
HMS *Vivid*
Service number: Z/8453
Age: 19

Buried at Sandwick.

Ordinary Seaman Donald MacLeod
10 Murray's Court, Stornoway
Son of John and Mary Ann MacLeod
Service: Royal Naval Volunteer Reserve,
RN Depot Crystal Palace
Service number: Z/9964
Age: 18

Buried at Sandwick.

**Engine Room Artificer 4th Class William
Kirk Wilson**
Beach House, South Beach, Stornoway
Husband to Mary Ann Nicolson (12 Gravir)
Service: RN, HMS *Mistletoe*
Service number: M/14184
Age: 29

Body never recovered.

Deckhand Alexander MacIver
42 Church Street, Stornoway
Service: Royal Naval Reserve,
HMD *Golden Effort*
Service number: 12120/DA
Age: 23

Survived.

Able Seaman Angus MacDonald
5 MacKenzie Buildings, Bayhead,
Stornoway (brought up at 32 Borve)
Service: Royal Naval Reserve,
HMS *Duchess of Devonshire*
Service number: 2712/A

Survived.

Leading Signalman Angus Nicolson
25 Battery Park, Stornoway
Service: Royal Naval Reserve,
HMS *Imperieuse*
Service number: 5136B
Age: 32
Survived. Died 5 July 1958, aged 73.

TOLSTA

Seaman John MacDonald
1 North Tolsta
Son of John and Marion MacDonald;
husband to Catherine
Service: Royal Naval Reserve,
HMS *Emperor of India*
Service number: 3339/B
Age: 40
Buried at Tolsta.

Deckhand Donald MacLeod
3 North Tolsta
Son of Donald and Ann MacLeod
Service: Royal Naval Reserve, HMS *Victory*
Service number: 3968/SD
Age: 20
Buried at Tolsta.

Seaman John Morrison
8 North Tolsta
Son of Kenneth and Mary Morrison
Service: Royal Naval Reserve
Service number: 4645/A
Age: 26
Buried at Tolsta.

Seaman John MacIver
33 North Tolsta
Son of John and Isabella MacIver;
husband to Dolina
Service: Royal Naval Reserve,
HMT *Letterflourie*
Service number: 2496/A
Age: 31
Buried at Tolsta.

Deckhand Donald MacIver
38 North Tolsta
Son of Kenneth and Annie MacIver
Service: Royal Naval Reserve, HMT *Iranian*
Service number: 18720/DA
Age: 26
Buried at Tolsta.

Seaman Donald Campbell
44 North Tolsta
Son of Hector and Catherine Campbell;
husband to Catherine
Service: Royal Naval Reserve, HMS *Victory*
Service number: 3356/B
Age: 47
Body never recovered.

Seaman Evander Murray
45 North Tolsta
Son of John and Margaret Murray;
husband to Margaret
Service: Royal Naval Reserve, HMS *Thames*
Service number: 1829/D
Age: 45
Buried at Tolsta.

Seaman Kenneth Campbell
54 North Tolsta
Son of John and Isabella Campbell;
husband to Mary Murray
Service: Royal Naval Reserve, HMS *Galatea*
Service number: 4804/A
Age: 29
Buried at Tolsta.

Seaman Donald MacLeod
58 North Tolsta
Son of Malcolm and Mary MacLeod
Service: Royal Naval Reserve
Service number: 3329/A
Age: 31
Buried at Tolsta.

Leading Seaman Malcolm MacLeod
58 North Tolsta
Son of Malcolm and Mary MacLeod
Service: Royal Naval Reserve
Service number: 5478/A
Age: 25
Buried at Tolsta.

Seaman John MacIver
69 North Tolsta
Son of Murdo and Isabella MacIver;
husband to Catherine
Service: Royal Naval Reserve, HMS *Victory*
Service number: 2619/C
Age: 48
Body never recovered.

Leading Seaman Murdo MacDonald
1 North Tolsta
Service: Royal Naval Reserve,
HMS *Emperor of India*
Service number: 3971/A
Survived. Died 11 February 1950, aged 61.

Seaman John MacInnes
2 North Tolsta
Service: Royal Naval Reserve,
HMS *Slains Castle*
Service number: 3403/B
Age: 32
Survived. Later moved to Gress.

Seaman Roderick MacDonald
23 North Tolsta
Service: Royal Naval Reserve,
HMS *Pembroke*
Service number: 6632/A
Survived.

Able Seaman Donald Murray
37 North Tolsta
Service: Royal Naval Reserve, HMS
Venerable; served in Royal Naval Division
in France, Flanders and Gallipoli; among
first on scene of torpedoed liner *Lusitania*,
May 1915.
Service number: 5575
Age: 23
Survived. Died 3 May 1992, aged 96.

Deckhand Donald MacIver
14 New Tolsta
Service: Royal Naval Reserve,
HMS *Joseph Burgin*
Service number: 19616/DA
Age: 18
*Survived. Married Christina MacLeod; they
settled in Inverness. Died 20 April 1986, aged
86.*

PARISH OF UIG

ISLE OF BERNERA

BREACLETE

Seaman John MacKenzie
8 Breaclete
Son of Malcolm and Annie MacKenzie;
husband to Mary
Service: Royal Naval Reserve, HMS *Victory*
Service number: 2038/C
Age: 39
Body never recovered.

HACKLET

Deckhand Donald MacAulay
4 Hacklet
Son of John and Christina MacAulay
Service: Royal Naval Reserve,
HMT *Max Pemberton*
Service number: 2658/SD
Age: 21
Buried at Bosta.

TOBSON

Leading Seaman Donald MacDonald
13 Tobson
Son of Angus and Mary MacDonald;
husband to Catherine MacDonald
Service: Royal Naval Reserve, HMS *Ganges*
Service number: 2688/B
Age: 36
Body never recovered.

Deckhand John MacKenzie
8 Tobson, Great Bernera
Service: Royal Naval Reserve,
HMT *Concord II*
Survived.

MAINLAND UIG

AIRD UIG

Malcolm MacRitchie
Last address in Lewis: 3 Aird,Uig
Service: Royal Naval Reserve, SS *Zaria*
Service number: 3312/B
Age: 39
Survived. Died 1962, aged 83.

BRENISH

Deckhand Murdo MacKinnon
18 Brenish
Son of Cain and Christina MacKinnon
Service: Royal Naval Reserve, HMD *Eddy*
Service number: 212090/DA
Age: 18
Buried at Ardroil.

Deckhand George Morrison
20 Brenish
Son of John and Marion Morrison
Service: Royal Naval Reserve, HMML 307
Service number: 3499/SD
Age: 20
Buried at Ardroil.

CLIFF

William MacLennan
36 Cliff
Service: Royal Naval Reserve
Age: 18
Survived. Died 8 January 1948, aged 47.

CROWLISTA

Seaman Angus MacDonald
6 Crowlista
Son of John and Annie MacDonald
Service: Royal Naval Reserve, HMS *Kent*
Service number: 4554/A
Age: 31
Buried at Ardroil.

Deckhand Murdo Nicolson
12 Crowlista
Son of Donald and Catherine Nicolson
Service: Royal Naval Reserve, Mercantile
Marine
Service number: 4258/SD
Age: 22
Body never recovered.

Deckhand Ewen MacDonald
13 Crowlista
Natural son of Kirsty MacDonald
Service: Royal Naval Reserve,
HMS *John Gray*
Service number: 19896/DA
Age: 18
Buried at Ardroil.

Seaman Malcolm MacKay
14 Crowlista
Son of Donald and Margaret MacKay
Service: Royal Naval Reserve,
HMS *Pembroke*
Service number: 7699/A
Age: 27
Buried at Ardroil.

Leading Seaman John MacDonald
16 Crowlista
Son of Angus and Flora MacDonald;
husband to Rachel
Service: Royal Naval Reserve,
HMS *Arrogant*
Service number: 2554/C
Age: 45
Body never recovered.

Seaman Peter Buchanan
23 Crowlista
Son of Donald and Chirsty Buchanan
Service: Royal Naval Reserve,
HMS *Pembroke*
Service number: 7700/A
Age: 32
Body never recovered.

EARSHADER

Seaman Kenneth Smith
1 Earshader
Son of John and Marion Smith; husband to
Christina MacKay of Crulivig (1888–1980:
last *Iolaire* widow)
Service: Royal Naval Reserve, HMS *Cove*
Service number: 1620/C
Age: 45
Buried at Bosta.

KNEEP

Able Seaman John MacLennan
15 Kneep
Son of John and Henrietta MacLennan, *née*
MacDonald
Service: Royal Naval Reserve, HMY *Iolaire*
(the original Stornoway depot-ship, built
in 1902 and transferred to Firth of Clyde,
October 1918)
Service number: HPT124Y
Age: 22

*Later emigrated to Australia, married
Margaret MacRitchie of 4 Aird Uig, returned*

*to Lewis c. 1935 and settled at that address.
One of the final handful of survivors, John
MacLennan died on 17 December 1987, aged
ninety-one.*

UIGEN

Deckhand John MacLeod
17 Uigen
Son of Murdo and Ann MacLeod
Service: Royal Naval Reserve, HMML 502
Service number: 44351/SD

Age: 22

Buried at Valtos.

Deckhand Angus Matheson
18 Uigen
Son of Malcolm and Catherine Matheson
Service: Royal Naval Reserve, HMD *Winner*
Service number: 18694/DA
Age: 19

Buried at Valtos.

PARISH OF HARRIS

ISLE OF BERNERAY

Seaman Norman MacKillop
Borve, Berneray
Natural son of Mary, by Neil MacKillop
Service: Royal Naval Reserve,
HMS *Pembroke*
Service number: 9522/A
Age: 19

Body never recovered.

Seaman Donald Paterson
Borve, Berneray
Natural son of Rachel MacLeod, by John
Paterson
Service: Royal Naval Reserve,
HMS *Pembroke*
Service number: 9521/A
Age: 18

Buried at Berneray.

North Harris

Deckhand Finlay MacLennan
Meavaig, Harris
Son of Christina MacLennan.
Service: Royal Naval Reserve
Service number: 18771/DA
Age: 20
Buried at Luskentyre.

Seaman Farquhar Morrison
Scrott, Stockinish
Son of Donald and Rachel Morrison;
widower of Rachel MacKinnon
Service: Royal Naval Reserve
Service number: 3161/B
Age: 35
Buried at Luskentyre.

Able Seaman Alexander Campbell
Myrtle, Plocropool
Service: RN, HMS *Royal Sovereign*
Age: 26
Survived. Served throughout Second World War; died 1972. His sister, Marion Campbell BEM (c. 1909–96), was a celebrated weaver.

Ordinary Seaman John MacKinnon
Smith Cottage, Tarbert
Service: RN, HMS *Edgar*
Survived. Died 1927.

Able Seaman Robert MacKinnon
Caw, Tarbert
Service: RN, HMS *Dublin*
Survived.

South Harris

Deckhand Kenneth MacLean
12 Northton,
Son of Donald and Mary MacLean
Service: Royal Naval Reserve,
HMS *Venerable*
Service number: 20994/DA
Age: 18
Buried at Scarista.

Seaman Norman MacKay
Obbe
Son of Neil and Flora MacKay
Service: Royal Naval Reserve,
HMS *Pembroke*
Service number: 9494/A
Age: 18
Body never recovered.

Seaman John MacCuish
Northton
Service: RN, HMD *Duthias*
Service number: 5755
Survived.

Isle of Scalpay

Deckhand Finlay Morrison
15 Scalpay
Son of Donald and Mary Morrison
Service: Royal Naval Reserve, HMML 560
Service number: 4515/SD
Age: 25
Buried at Luskentyre.

Other Long Island Survivors of HMY *Iolaire* (Address Unknown)

Seaman John MacDonald
Service: Royal Naval Reserve,
HMS *Pembroke*
Service number: 4534A
Survived.

Deckhand Murdo MacKay
Service: Royal Naval Reserve, HMS *Tarlair*
Service number: 18856/DA
Survived.

Seaman Murdo MacLeod
Service: Royal Naval Reserve, Adams
Depot, Middlesborough
Service number: 4154
Survived.

Seaman Murdo MacLeod
Service: Royal Naval Reserve, HMS *Grouse*
Service number: 3316/C

Survived.

Deckhand Norman MacLeod
Service: Royal Naval Reserve, HMML 214
Service number: 3556/SD

Survived.

Stoker George Murray
Service: RN, HMS *Dominion*
Service number: 2789/K

Survived.

Total on Board: 284
Total Lost Listed Here: 201
Total Saved: 80

Total Lost Reported in 1919: 205,
from which 205,
– Lewis casualties: 178
– Harris casualties: 7
– Officers and crew: 20

Number of bodies never recovered: 56

Number of bodies recovered but never identified: 8 (buried in a mass grave at Sandwick, with three named casualties)

'Unrecovered' total: 64

NOTE

I am indebted to Bill Lawson for devoting two full days, amidst many duties and demands, to help with this list and nail down much important detail; and especially our respected local historian, Malcolm MacDonald. I have also been assisted or corrected by some individuals, like Ruairidh Moir of Tolsta and Mòr MacLeod of Brue, and on occasion have resorted to visiting a distant cemetery to check some detail against the headstone.

In 1919 the Admiralty counted 205 lost and 79 saved. An eightieth survivor, Neil MacKenzie of 14 Leurbost, was recently named by Malcolm MacDonald, who has to date found hard details for 201 dead. Assuming the 1919 figure was accurate (bearing in mind the real risk of duplication between common island names and shifts of addresses) several more may yet be listed.

In some instances, those in search of an *Iolaire* grave should note, families eschewed an official Imperial War Graves Commission headstone to erect one of their own. For example, private memorials mark the rest of Norman MacLeod of 13 Arnol (at Bragar cemetery) and Kenneth Smith of 1 Earshader (at Bosta). Inevitably there are contradictions. The ages given for several Crowlista casualties on the Uig War Memorial vary, by a year either way, from Mr MacDonald's list. In a surprising number of instances, too – especially those whose bodies were never recovered and whose deaths were registered late in 1919 – the age then given is at variance with their entry in the civil register of births. (In all these instances, where the disparity is known, I have calculated from the birth year.) This is a problem even in relatively recent times – into an age of State benefits when, for the first time, the civil record became practically important for ordinary people. Bill Lawson had once to supply evidence to support the claim of a spry island lady who, due to the incompetence of one Victorian registrar on Lewis, had never legally been born at all.

No one will ever compose a list of passengers and crew beyond dispute in every detail. No record was taken at Kyle on 31 December 1918 – there is nothing sinister in this; it is less than a decade since Caledonian MacBrayne began to keep a detailed passenger-list, and that only on the longest sailings – and there are additional complications as to addresses within Lewis. In the course of a war that lasted four years, men often did move, especially in the overcrowded Lewis conditions of the time; and many married during the conflict and flitted on that basis. Like confusion, posthumous births and the depredations of the 1918–19 influenza pandemic (to say nothing of high neonatal and infant mortality generally) also preclude an absolute number of widows and orphans, though 67 and 209 are accepted as roughly accurate. It is especially difficult, too, to gain details of age and death for those survivors who emigrated overseas and never returned. It was also by no means uncommon for men to lie about their age – boys exaggerating upwards, fifty-something men downwards – to gain or retain RNR service. Not all who stayed on Lewis, or who returned to it – and were duly interred there – lie in marked graves; nor were they always buried in their native district. Island headstones, besides, do not always give the local address – an important detail when you are confronted with perhaps a dozen Donald MacLeods of the appropriate age in a local burial-ground – and of all who boarded the *Iolaire* and lived to tell the tale, only the gravestone of Donald Morrison, *Am Patch*, refers to the disaster.

At the death, on 27 June 1992, of Neil Nicolson – officially the last survivor – there were strong rumours of another still thought to be alive in Los Angeles, but no one seemed quite sure who, or where he was from, and it is most improbable the mystery sailor is alive now.

Six pairs of brothers died in the *Iolaire* disaster – the MacLeans of 6 South Bragar, the MacDonalds of 13 Swainbost, the MacLeods of 58 North Tolsta, the Campbells of Vatisker (Back), the two John MacDonalds of 20 Sheshader, and the MacKenzies of 16 Leurbost. It was by no means uncommon for siblings to have the same given name, and this endured locally till very recently, there being a strict order of naming successive offspring in honour of senior relatives; and three, four or even five Donalds (or Murdos, or Anguses) is by no means unheard of within the same family.

All additional information, corrections or suggested adjustments to what must remain a work in progress would be welcomed.

John MacLeod
jm.macleod@btinternet.com

Sources

BOOKS ON HMY *IOLAIRE*

Loyal Lewis Roll of Honour, ed. William Grant (2nd edn., 1920). An astonishing record of the island's contribution to the Great War, with portraits of many who gave their lives, including those lost with the *Iolaire*. Unfortunately, this haunting work is extremely rare. An earlier edition was printed in 1915.

Sea Sorrow: The Story of the Iolaire Disaster, ed. James Shaw Grant (Stornoway Gazette, c. 1952; new edition with additional material 1972; reprinted c. 1989). A 40-page booklet consisting largely of 1919 reportage from the *Stornoway Gazette*, including accurate, if not quite complete, casualty list, verbatim account of the Fatal Accident Inquiry and some English poems.

Call na h-Iolaire [*The Loss of the* Iolaire], Tormod Calum Dòmhnallach (Acair, Stornoway, 1978). A 124-page paperback, largely in Gaelic, with a 14-page English synopsis and a 'Navigational Appendix' by Alexander Reid. There are several clearly drawn charts and evocative, largely pastoral photographs.

An Cogadh Mór 1914–1918, ed. Andrew Moir (Acair, Stornoway, 1982). An engaging Gaelic booklet, thick with evocative contemporary photographs, of the Hebridean contribution to the conflict, drawing heavily on interview material with elderly veterans – including Donald Morrison, *Am Patch*, and his astounding escape from the *Iolaire*.

Several inadequate accounts of the tragedy have appeared in general volumes of Scottish or maritime woe. See the chapter entitled 'The Brave

That Are No More' in *Scottish Disasters*, Donald M. Fraser (Mercat Press, Edinburgh, 1996); and the appalling 'HMS *Iolaire* – 1918' in *SOS: Men Against The Sea*, Bernard Edwards (New Guild, Dorset, 2002), the woes of 'Clement Cotter' and his command being detailed largely from the author's rich imagination.

For the first published memoir of a survivor, see Telegraphist Leonard Welch's heavily embroidered version of events in 'Wrecked on the Beasts of Holm' in a 1942 anthology, *My Amazing Adventure: A Collection of True Stories* (T.V. Boardman and Co., London and New York), pp. 12–18. A 1956 account, originally published in an unnamed Canadian magazine and subsequently lifted by the *Stornoway Gazette* that summer, is anonymous and, from internal evidence, should be treated with profound suspicion.

The present century has brought a highly ambitious romantic novel set in the Stornoway of the Great War and pegged to the disaster – *The Dark Ship*, Anne MacLeod (11:9 [Neil Wilson Publishing], Glasgow, 2001). Though the book is tainted by prevalent modern attitudes to religion, relationships and the First World War, the set-piece description of the wreck is excellent.

At the time of first writing – October 2008 – a book on the disaster was understood to be in preparation by Malcolm MacDonald and Donald J. MacLeod. As of January 2013, the fruits of this work had yet to appear in print.

BOOKS ABOUT THE GREAT WAR

As the First World War pours over the edge of living memory, there has been substantial reassessment of the conflict and a new, robust school of 'revisionist' history.

The best, detailed, one-volume account, scrupulously researched, is *1914–1918 – The History of the First World War*, David Stevenson (Allen Lane, London, 2004).

The Great War, Correlli Barnett (Park Lane Press, 1979; republished by the BBC, 2003). A very clear and accessible narrative: Barnett was co-scriptwriter on a landmark 1964 BBC television documentary.

Mud, Blood and Poppycock, Gordon Corrigan (Cassell, London, 2003). Though to some extent indebted to Sheffield, if still more assertive in myth-busting, this is a robust correction to the popular picture of the war on the Western Front, given added steel by Corrigan's own experience as a regular officer of the Royal Gurkha Rifles.

The Pity of War, Niall Ferguson (Allen Lane, London, 1998). Though at times very worthy, this is an important and sobering book.

The Flowers of the Forest: Scotland and the First World War, Trevor Royle (Birlinn, Edinburgh, 2006). A masterly study, with two pages on the loss of the *Iolaire*.

Forgotten Victory – The First World War: Myths and Realities, Gary Sheffield (Headline Book Publishing, London, 2001). Pungent, iconoclastic and powerful.

From Sarajevo to Postdam, A.J.P. Taylor (Thames and Hudson, London, 1966). A concise, illustrated and opinionated survey of European history, political and cultural, from 1914 to 1945 – guaranteed to stimulate.

The Great War – Myth and Memory, Dan Todman (Hambledon Continuum, London, 2005). A deft exploration of prevalent misconceptions of the First World War, and their origins in the bitterness of the inter-war decades.

Books about Lewis and Harris

Gleanings of Highland Harvest, Rev. Murdoch Campbell (2nd edn, Ross-shire Printing and Publishing Co., 1957). There is also an earlier, 1953, edition as well as a 1989 paperback by Christian Focus Publications Ltd, Fearn, with new introduction by Rev. J.D. MacMillan. A collection of anecdotes and 'testimonies' from the Highland Evangelical tradition, including material on North Tolsta.

Ness Cemetery Records 1916–1994 (Comann Eachdraidh Nis, Habost, Ness, Isle of Lewis, 1995). At first glance a volume of extraordinary morbidity – an exhaustive log of local burials, with stark John MacKinnon cover-photograph and several apt texts throughout the book from the Authorised Version. In fact, it is an absorbing read and attests vividly to high child mortality in the early twentieth century and especially in

1919. Comann Eachdraidh Nis – the local history society – issues new, updating pages for insert every year.

Children of the Black House, Calum Ferguson (Birlinn, Edinburgh, 2003), with foreword by Donald Meek. A superb account of Lewis life from the middle of the nineteenth century, as experienced by one family in the Point district and carried down in their tradition. See pp. 160–171 for a searing description of the *Iolaire* tragedy.

Lewis In The Passing, ed. Calum Ferguson (Birlinn, Edinburgh, 2007). A powerful collection of interviews with older Lewis people, recording their experiences of the twentieth century, with some reference to the Great War and the *Iolaire*.

Highland Villages, James Shaw Grant (Robert Hale, London, 1977) – *Iolaire* references pp. 22, 23.

The Hub of My Universe, James Shaw Grant (James Thin, Edinburgh, 1982); *Surprise Island* (James Thin, Edinburgh, 1983). The first volumes in a series of anthologies of Grant's long-running 'In Search of Lewis' *Stornoway Gazette* column, from 1980 to 1999: his very last piece, that January, was a conscientious essay by a tired old man as to why the paper had for eighty years ignored the *Harris* casualties of the wreck. These early columns include several meditations on the *Iolaire*.

Discovering Lewis and Harris, James Shaw Grant (John Donald Publishers, Edinburgh, 1987); *Iolaire* material pp. 50–54.

Clar-Ainm Urramach Na Hearadh [*Harris Roll of Honour 1939–1945*], Harris Council of Social Service, with foreword by Rev. Professor Murdo Ewen MacDonald (Stornoway Gazette, 1992). A detailed record, with many photographs, of the contribution of Harris villages to the defeat of Nazi Germany.

The Soap Man – Lewis, Harris and Lord Leverhulme, Roger Hutchinson (Birlinn, Edinburgh, 2003). An excellent analysis of Lord Leverhulme and his failure on the Long Island, balancing and in some respects correcting the much more regretful *Lord of the Isles*, Nigel Nicolson (Weidenfeld and Nicolson, London, 1960); new paperback edition of the Nicolson text published by Acair. (Stornoway, 2000), with a hideous cover.

Sgeulachdan a Seisiadar – Tales from Sheshader, Chris and Bill

Lawson (Bill Lawson Publications, Northton, Isle of Harris, 1990). A vivid, engaging collection of anecdotes from the life of one Lewis township and – with each in parallel English translation – an excellent tutorial for the Gaelic learner. Sheshader was one of the villages most devastated by the calamity of 1 January 1919: that there is here only one, passing, reference to the *Iolaire* subconsciously attests to the depth of that trauma.

Dateline: Stornoway, Bill Lucas (privately published, 2008). This memoir of five decades in local journalism has some interesting *Iolaire* material.

Croft History, Isle of Lewis, vol. 15, Gleann Tholastaidh, Bill Lawson (Bill Lawson Publications, Northton, Harris, 2008).

Croft History, Isle of Lewis, vol. 15, Tolastadh – Am Baile Ur, Bill Lawson (Bill Lawson Publications, Northton, Harris, 2008). A meticulous croft and family history of the community.

A Lewis Album: From the Collection of Historical Photographs compiled by Angus M. MacDonald ARPS, ed. Sheila MacLeod (Acair, Stornoway, 1982). Many excellent images of island life from late Victorian times.

Lewis: A History of the Island, Donald MacDonald (Gordon Wright Publishing, Edinburgh, 1978), with preface by the Rt. Hon. Donald Stewart MP. Though it would have been none the worse for judicious editing, and besides arranged chronologically rather than thematically, it remains the best study.

The Tolsta Townships, Donald MacDonald (North Tolsta Historical Society, 1984). An outstanding description, rich with anecdote and vivid descriptive material. Unfortunately, it is all but unobtainable. There are hopes of republication.

Pilgrim Souls, Mary and Hector MacIver (Aberdeen University Press, 1990). Includes worthwhile autobiographical material written or broadcast by Hector MacIver (1910–66), best remembered as a celebrated teacher of English at the Royal High School, Edinburgh. His pupils included the late Robin Cook, and he was a popular figure in the city's literary circles, memorialised in the festschrift *Memoirs of A Modern Scotland*, ed. Karl Miller (Faber & Faber, London, 1970), with contributions from 'Hugh MacDiarmaid', Sorley MacLean, George Mackay Brown, William

McIlvanney, etc. MacIver's life, in all its sparkling ability, was tragically shortened by alcohol-related illness.

Devil in the Wind, Charles MacLeod (Gordon Wright Publishing, Edinburgh, 1979). An autobiography of 1950s life on the West Side of Lewis, most lightly disguised as a novel, deftly combining harrowing village folklore of the Great War, social injustice and other hard times with a jollier present – a useful corrective to the prevalent mythology that the events of 1 January 1919 plunged Lewis into decades of unrelieved gloom.

Survival Against The Odds – The Story of Petty Officer Donald MacKinnon: Russian Convoy Survivor, Donald J. MacLeod (Loch Roag Books, Aberdeen, 2000). MacLeod is a passionate student of Long Island seamanship and its largely unheralded part in two global conflicts, and this is an absorbing little book with some intense reference to the *Iolaire*.

Nis Aosmhor. The Photographs of Dan Morrison (Acair Ltd., Stornoway, 1997), with Gaelic introduction by Dr Finlay MacLeod and English foreword by Rev. Professor Donald Macleod. The fine selection of photographs of community life through half a century by Dan Morrison of Habost (1912–2009) includes a masterly, 1963 study of two *Iolaire* heroes, Donald Morrison and John Finlay MacLeod, working on an oar at the MacLeod boatyard; and two striking portraits of Mr Morrison's mother, Jessie Morrison or Gunn, who lived till 1961 and was still exuberantly driving cattle in her nineties.

One Man's Lewis – A Lively View of a Lively Island by The Breve, George Morrison (Stornoway Gazette Ltd., c. 1979), with foreword by J.S. Grant. A delicious collection by a sparkling local columnist.

Lewis and Harris Seamen 1939–1945, John Morrison and Annie Morrison (Stornoway Gazette, 1993), with foreword by James Shaw Grant. A gripping narrative of Long Island engagement in Hitler's war, unfolding chronologically and splendidly illustrated. Mr Morrison, a son of North Tolsta who passed away in October 2008, had a distinguished police career: he commanded the successful manhunt for Donald Neilson, the 'Black Panther', and retired as Deputy Assistant Commissioner of the Metropolitan Police

Roll of Honour – Ness to Bernera – For King and Country 1939 to 1945, ed. Ruaraidh Moireasdan (Stornoway Gazette, 1988), with foreword by Rev. Murdo MacAulay. Another impressive record of names and photographs. The book concludes with a postscript by two of the last HMY *Iolaire* survivors, Donald Morrison and John MacLennan.

The Islands of Western Scotland – The Inner and Outer Hebrides, W.H. Murray (Eyre Methuen Ltd., London, 1973). The finest book about the Hebrides ever written.

The Voice of the Bard – Living Poets and Ancient Tradition in the Highlands and Islands of Scotland, Timothy Neat with John MacInnes (Canongate, Edinburgh, 1999). See profile of Kenneth J. Smith, Earshader, Isle of Lewis, pp. 94–107, with anecdotes of the *Iolaire* and the loss of his father.

Carn Cuimhneachain: Memorial Cairn, North Tolsta Community Council, prepared by Angus MacLeod. A meticulous and beautifully presented souvenir list of all the village men who gave their lives in the wars of the twentieth century.

Around the Peat Fire, Calum Smith (Birlinn, Edinburgh, 2001). Another memoir of Lewis youth and Shawbost boyhood, with some original material on the *Iolaire*, including the valuable recollections of Lieutenant Townend on his grim organisation of an impromptu mortuary.

Harris and Lewis – Outer Hebrides, Francis Thompson (2nd edn., David and Charles Ltd, Newton Abbot, 1973). A fine and thoughtful analysis of the Long Island with passionate reference to the *Iolaire* disaster. Other editions were printed in 1968 and 1987.

Metagama: A Journey from Lewis to the New World, Jim Wilkie (Mainstream Publishing Co. [Edinburgh], Edinburgh, 1987). This has recently been reprinted in paperback by Birlinn. (Edinburgh, 2001). A tight, excellent account of 1920s emigration.

RELEVANT BOOKS ON MATTERS MARITIME

Na Nuadh Bhataichean: The MacBrayne Fleet 1928–1964, Ailean Boyd (Acair, Stornoway, 2006). An evocative, bilingual account of fondly remembered MacBrayne motor vessels with, especially, an interesting profile of the 1947 *Loch Seaforth* – probably the most cherished ship ever to serve Stornoway – and her sticky end at Tiree in March 1973.

Nicholl's Seamanship and Practical Knowledge, for Second Mates', Mates' and Masters' Examinations, Charles H. Brown FRSGS (19th edn., Brown, Son and Ferguson, Glasgow, 1952). The sort of stolid textbook by which British marine officers won two world wars.

West Highland Steamers, C.L.D. Duckworth and G.E. Langmuir (4th edn., Brown, Son and Ferguson, Glasgow, 1987). A classic of meticulous detail and tight, at times most elegant, writing, with a very good account of the 1904 *Sheila* and her own wreck at Cuaig.

Mayday: The Perils of the Wave, Nicholas Faith (Channel 4 Books, 1998). Accompanying a TV series that was not unduly sensational, this is an accessible but sober account of disaster at sea and what remains an alarmingly careless, unregulated global industry.

The Riddle of the Titanic, Robin Gardiner and Dan van der Vat (Weidenfeld and Nicolson, 1995). Rightly described in *The Times* as a 'well-written, well-illustrated and intellectually satisfying study of the world's most celebrated sinking'; and an object-lesson in how good writers should handle chaotic, often conflicting testimony.

Titanic: The Ship That Never Sank, Robin Gardiner (Ian Allan Publishing Ltd, Shepperton, 1998). Given free reign to pursue a fantastic conspiracy theory, this Gardiner outing is less masterful than his earlier collaboration, and at times jarringly self-indulgent, but nevertheless raises interesting questions – to say nothing of dark material on death by immersion.

The Yachtsman's Pilot to the Western Isles, Martin Lawrence (Imray, Laurie, Naurie and Wilson, Huntingdon, 1996). An accessible guide to the navigation of, amid much else, Stornoway Harbour, which may now be a little dated. A professional seaman would no doubt prefer the relevant and rather beautiful Admiralty chart, *Approaches to Stornoway*, 'published at Taunton, United Kingdom 14th April 1978 under the Superintendence of Rear-Admiral D.W. Haslam OBE, Hydrographer of the Navy.' A copy purchased at the Shipping Services Office in Stornoway in September 2008 cost £20, was the November 1997 edition, and had been updated by hand with a few new soundings and temporary hazards only this year.

It Must Be Stornoway: The Story of Stornoway Pier and Harbour Commission 1865 to 2004, Catherine MacKay (Argyll Publishing, Glendaruel, 2008). An excellent read with many photographs and reproduced ephemera. Harbour administration passed in May 2004 to a new body, Stornoway Port Authority, under the terms of recent legislation.

The North Herring Fishing: Ring Netting in the Minches, Angus Martin (House of Lochar Publishers, 2001). An 'oral history' of stories Martin collected from Kintyre and Ayrshire fishermen, whose depredations wrought such havoc on the Hebridean economy between the wars and provoked Compton MacKenzie and John Lorne Campbell to establish the Western Isles Sea League in defence of island interests. But amidst the remorseless narrative of rapacity is a gem of a story about James MacLean's first visit to Stornoway since the *Iolaire* was wrecked, in 1948, as told by Duncan MacArthur – his grandson – to Angus Martin.

The Kingdom of MacBrayne, Nick S. Robins and Donald E. Meek (Birlinn, Edinburgh, 2006). A lavishly illustrated history of David MacBrayne Ltd, with some reference to modern car ferry developments; though with undue emphasis on the Inner Hebrides and especially Tiree. A paperback edition is now available.

POETRY

An Toinneamh Diomhair – Na h-Orain aig Murchadh MacPhàrlain, Bàrd Mhealabost ['*The Secret Spinning – Songs by Murdo MacFarlane, Bard of Melbost*'] (Stornoway Gazette, c. 1972). MacFarlane (1901–82) was a warm and sparkling writer and a compelling man who wrote not only original Gaelic lyrics but, most unusually, original airs for them. This collection includes two poems on the *Iolaire* disaster, of which *Raoir Reubadh an Iolaire* – popularly known as *Oran an Iolaire* – is decidedly the better. A noted recording by Catherine Anne MacPhee, to MacFarlane's own melody, has won it a huge new audience. The permission of Mrs M.F. Campbell, Tong, to publish the poem is gratefully acknowledged.

Iain Crichton Smith wrote two poems entitled 'The *Iolaire*' and seems to have grown disaffected with this 1957 essay; it is pointedly excluded from the Carcanet edition of his *Collected Poems*, 1992. That is a great pity: it is much superior to the later poem, inaccessible and dark, putting himself in the mind of a church elder trying (less than successfully) to comprehend the horror within his belief system. Born in Glasgow in 1928, and brought up in Bayble, Point. Smith was a gifted writer in both Gaelic and English, marked by concision, droll wit, an uneasy relationship with religion and great versatility of genre, producing novels, short stories and plays as well as outstanding poetry. He died in 1998. His original *Iolaire* poem was published in the Winter 1957 number of the *Saltire Review*, a short-lived Scottish literary quarterly. The poet's widow, Mrs Donalda Henderson, has kindly allowed its appearance here.

Rev. John MacLeod was born at 13 Arnol in January 1918 and would have no memory of his father, Norman, drowned with the *Iolaire* the following New Year. His mother, Catherine, never recovered, and at her death in 1933 an aunt returned from Canada to look after her two boys for a year. He spent three years in Royal Navy service during the Second World War before completing his training for the Church of Scotland ministry, being ordained as a naval chaplain in May 1945. He held pastorates in Canada, Edinburgh and Oban, retiring in 1983, and for many years contributed Gaelic poems and articles to assorted publications; probably his best known Gaelic song is '*Nam Aonar le Mo Smaointean*, on the Falklands War of 1982, covered by Anne Lorne Gillies, Alasdair Gillies and most successfully by James Graham. A respected scholar and an outstanding Christian, MacLeod died in December 1995. 'Bantrach Cogaidh' – 'War Widow' – seems first to have been published in *Gairm* (no. 85) in 1973 and then quoted in Dòmhnallach's *Call na h-Iolaire* in 1978, but may have been written many years earlier; MacLeod was evidently haunted by memories of his mother's obsessive washing of his father's ripped and threadbare Navy uniform every summer till the end of her life. The poem, in Ronald Black's fine translation and with his brief account of MacLeod's life, can be found in *An Tuil: Anthology of 20th Century Scottish Gaelic Verse*, ed. Ronald I.M. Black (Polygon, Edinburgh, 1999). The permission of the late Mrs Doleena MacLeod, Oban, to publish her late husband's poem is gratefully acknowledged.

There are many other poems and songs about the *Iolaire*. Most are dreadful.

Court of Inquiry, etc., into the loss of HMY Iolaire, 1st January 1919. ADM 116/1869 – formerly File 693, the National Archives (Public Records Office), Kew, London. This includes telegrams, correspondence, witness depositions, stenographed testimony of survivors, and assorted memoranda. Some material referred to within such to-and-fro is evidently missing. Under the 'Fifty Year Rule', File 693 was not made available to the public until New Year 1970. A copy is held at Stornoway Public Library.

The Fatal Accident Inquiry of February 1916 was covered in detail by the Press – File 693 includes *The Scotsman*'s report of proceedings – and an entire verbatim transcript; the *Stornoway Gazette* account can be perused on microfilm at Stornoway Public Library. Practically everything is reprinted verbatim in the *Sea Sorrow* booklet.

The papers of the Iolaire Disaster Fund are held in store by Museum nan Eilean, Francis Street, Stornoway, at their depository in Marybank, and may be viewed by appointment with Museum officials. These are fragile documents which will not bear much more handling, and it is probable a photocopied or digital version will be prepared for future scholarship.

Stornoway Public Library also hold a scant box-file of some relatively recent press-cuttings and articles on the *Iolaire* disaster, stored in the public reference area and which can be viewed without appointment.

The Angus MacLeod Archive is a remarkable collection of books, documents and tape-recordings on matters of South Lochs, Lewis, Highland, Gaelic and Evangelical religious interest amassed by the late Angus MacLeod MBE, as part of his effort to build a considerable resource of local history in his native South Lochs. It is now held at the Ravenspoint Centre, Kershader, South Lochs, Isle of Lewis. Much has been made available online – www.angusmacleodarchive.org.uk/ – and of immediate interest are recordings of two BBC programmes on the *Iolaire* disaster, which can be heard online or downloaded as an MP3 file. These are the 1959 Fred Macaulay programme, in Gaelic, and the soundtrack of the excellent *The Iolaire Disaster*, broadcast on BBC TV in 1974 as part of their 'Yesterday's Witness' series. Some footage was recycled for a disappointing BBC Alba documentary on New Year's Day 2009; by contrast a 1979 Grampian TV programme, *Home at Last*, though twenty years old and made on an evidently low budget, and repeated that night, was very much stronger.

Most of the local history societies on Lewis have collated papers and recordings on the *Iolaire* and two notable examples are in part available online: that of Comann Eachdraidh Nis, www.c-e-n.org/iolaire.htm; the excellent Tolsta website, www.tolsta.info/iolaire.htm; and the outstanding http://iolaire1919.blogspot.com/, prepared by Malcolm MacDonald of the Stornoway Historical Society, with a related page www.adb422006.com/iolaire.html. Much Comann Eachdraidh material is, of course, in Gaelic; most are run by unpaid volunteers and understanding of present copyright law tends to be uncertain and varies wildly.

A page of the ambitious Hebridean Connections website, www.hebrideanconnections.com/Details.aspx?subjectid=14576, bookmarks various items of interest from comparable organisations in Uig, Bernera and South Lochs, including material on John MacLennan (an important interview by Maggie Smith, in English translation) and Neil Nicolson.

Comann Eachdraidh Tholastaidh bho Thuath devoted much of the Winter 2006 number of their quarter publication, *Seanchas*, to the *Iolaire* disaster, with photographs and commentary; there have also been fine publications by Comann Eachdraidh Nis, including issues of their journal *Criomagan* – vol. 2, issue 1, January/February 1997; vol. 2, issue 3, March/April 1997, and especially vol. 4, issue 1, January/February 1999. Copies of all these can readily be bought from the respective societies, or viewed on their premises or at Stornoway Public Library.

The Stornoway Historical Society – and in particular the diligent Malcolm MacDonald – have done outstanding work, especially in compiling an exhaustive list of *Iolaire* casualties and survivors. This material, well illustrated, can be viewed online at http://iolaire1919.blogspot.com In 2007 the Society organised an important exhibition in the town, complete with a period diver's helmet, a model of the ship, a piece of her ensign, Great War medals, a 1919 chart of Stornoway harbour and many interesting photographs.

There are a surprising number of websites devoted to the *Iolaire*: others worth a look include Tony Wade's fine article on the Culture Hebrides page, www.culturehebrides.com/heritage/iolaire/ and a careful piece in the 'Scots at War' series, www.scotsatwar.org.uk/AZ/iolaire.htm. Some audiovisual material can be viewed on YouTube, notably a little film by 'JohnPaul33', www.youtube.com/watch?v=F6mrOm7g6ao

In liaison with the 2009 BBC Alba documentary, a majestic Gaelic multi-media website, with striking archive images and original, haunting drawings was unveiled for the ninetieth anniversary - http://www.bbc.co.uk/alba/tbh/iolaire/

In chill but beautiful, sunny conditions, a sombre service of remembrance was held by the monument at the Holm site on Thursday 1 January 2009, with a considerable attendance including contingents from all emergency services. Mor MacLeod and Alasdair 'Sandy Mor'MacLeod – the last known Iolaire orphans – were honoured guests. A similar ceremony was also conducted by the War Memorial in Tarbert, Harris.

ARTEFACTS

Various items are held at Museum nan Eilean, Francis Street, Stornoway; the ship's bell and engine-builder's plate are usually on permanent display, and a lifejacket (donated by the family of Donald MacIver after his death in 1986) and some small items are also stored there. A fragment of a Royal Navy ensign of the Great War, authenticated as such by Historic Scotland, is held by Stornoway Historical Society and the little scrap of red cloth, now framed, is said to be from that of the *Iolaire*. In dives on the site and since 2008, Chris Murray has recovered a number of important items and donated these to Museum nan Eilean. The most poignant are several buttons from an officer's jacket – and Lieutenant Cotter's Very pistol. Various other artefacts and papers from January 1919 are in private hands; a telegraphed summons to the Court of Inquiry received by Alick Campbell in Plocrapool, for instance, is still held by his family.

LOCALE

The monument at Holm Head is well signposted, off the main road at Sandwick between Stornoway and the island airport; the straight single-track road, replete with 'traffic-calming measures', serves two working farms and should be taken slowly. The last stretch is shut off by a locked gate in late evening and concludes with modest car-parking, a wicket-gate to access the footpath and an interpretative plaque. Most should be able to walk the short distance to the memorial and cairn and, the footpath apart, the scene is scarcely altered since 1919. The Holm coast is dangerous and care should be taken by the cliff edge; there is one treacherous little chasm very near the monument. There are no restrictions on diving at the site of the *Iolaire* wreck, but it calls for experience, excellent knowledge and proper support, and the seabed here should be treated with the same reverence as any other cemetery.

Acknowledgements

Donald Morrison, *Am Patch*, 7 Knockaird
(1898–1990)
Donald Murray, *Dòmhnall Brus*, 37 North Tolsta
(1895–1992)
Survivors, HMY *Iolaire*, 1 January 1919

Peggy Gillies, *née* Murray, born 36 Lionel
(1915–2006)
and
Mrs Marion MacLeod, *née* Smith, born 1 Earshader (1914–2012),
daughters of the *Iolaire* disaster

Mr Ruairidh Moir and Comann Eachdraidh Tholastaidh Bho Thuath;
North Tolsta Historical Society; Mr Norman Smith, *Tormod Sguigs*, Lionel;
Mr Malcom MacDonald, Stornoway Historical Society; Mr Mark
Elliot, Museum nan Eilean, Stornoway; Mrs Marina Davenhill and all
staff of Stornoway Public Library; Chris and Bill Lawson, Northton,
Harris; Staff of The Scottish Room, Edinburgh Central Library; The
National Archives, Kew; The Imperial War Museum, London; Mr John
J. MacLennan, Chief Executive, Stornoway Port Authority; Mr Andrew
Murray, great-grandson of Thomas Gusterson, North Tolsta; Mr Chris
Murray; Mr Jason King; Mr Teàrlach Quinnell; Hugh Andrew, Andrew
Simmons and all at Birlinn; an outstanding editor, Nancy E.M. Bailey;
and, for reading an early typescript and for many excellent suggestions,
Rev. Dr Donald Macleod DD; Mr Norman Campbell; Dr Robert Dickie
FRCGP DRCO; Mrs Mary Ferguson and Captain Angus M. MacKenzie
MN, Harbourmaster of Stornoway, 1978–2004